Digitalisierung im Einkauf

Florian Schupp · Heiko Wöhner
(Hrsg.)

Digitalisierung im Einkauf

Springer Gabler

Herausgeber
Florian Schupp
Bühl, Deutschland

Heiko Wöhner
Baden-Baden, Deutschland

ISBN 978-3-658-16908-4   ISBN 978-3-658-16909-1   (eBook)
https://doi.org/10.1007/978-3-658-16909-1

Die Deutsche Nationalbibliothek verzeichnet diese Publikation in der Deutschen Nationalbibliografie; detaillierte bibliografische Daten sind im Internet über http://dnb.d-nb.de abrufbar.

Gedruckt auf säurefreiem und chlorfrei gebleichtem Papier

Springer Gabler ist Teil von Springer Nature
Die eingetragene Gesellschaft ist Springer Fachmedien Wiesbaden GmbH
Die Anschrift der Gesellschaft ist: Abraham-Lincoln-Str. 46, 65189 Wiesbaden, Germany

Vorwort

Wer sich im Umfeld des Einkaufs bewegt, erlebt täglich Lieferprobleme, Bestellverfolgung, Lieferantensuche und Verhandlungsherausforderungen. Daneben kommen unvermittelt Qualitätsprobleme und Technologiefragen in der Zusammenarbeit mit Lieferanten auf.

Gleichzeitig sind Einkäufer wie auch Lieferanten mit neuen Trends konfrontiert, deren positive Elemente sofort in die tägliche Arbeit übernommen werden sollen. Ein großes Thema dabei, vielleicht das derzeit größte Thema, ist die Digitalisierung.

In diesem Spannungsfeld wollen wir zwischen dem großen Wort Digitalisierung und der gelebten Praxis im Einkauf eine Brücke schlagen und chancenorientiert eine fundierte und umsetzbare Entwicklung ermöglichen.

Dabei haben wir aus unserer Sicht sinnvolle Ansatzpunkte für die Digitalisierung gewählt und für die Untersuchung, Strukturierung und Entwicklung dieser Ansatzpunkte themenspezifisch ausgewiesene Experten für einen Beitrag zu diesem Buch angefragt. Einige dieser Experten leben oder arbeiten im nicht-deutschsprachigen Raum, weshalb wir deren Beiträge in englischer Sprache einbinden.

Allen Experten danken wir sehr herzlich für ihre Beiträge zu diesem Buch. Nur über diesen Weg war die thematische Bearbeitung sowohl in der Breite als auch in der Tiefe möglich. Außerdem gilt unser Dank Michael Hartig, Arnaud Molique, Michael Münig, Martina Lioi und Susanne Kramer.

Wir wollen unser Netzwerk erweitern und laden Sie deshalb zum Dialog im Themenbereich „Digitalisierung im Einkauf" ein.

Unser Engagement ist nur möglich mit der Unterstützung, dem Rückhalt und der Geduld unserer Familien. Daher gilt unser besonderer Dank Alexandra, Liv Grete, Anne-Hélène, Immanuel, Julia und Hinnerk.

Florian Schupp
Heiko Wöhner

Inhaltsverzeichnis

Ansatzpunkte für Digitalisierung im Gestaltungsbereich des Einkaufs

Florian Schupp und Heiko Wöhner

Zusammenfassung

Der Gestaltungsbereich des Einkaufs umfasst neben den klassischen Bereichen Lieferantensuche, Verhandlung und Nominierung auf der einen Seite auch die Bereiche der Lieferantenentwicklung in Bezug auf Qualität und Technologie sowie auf der anderen Seite das Supply Management mit dem Ziel der Sicherstellung einer effizienten Produktion und Produktionssteuerung. Diese Elemente des Einkaufs können in unterschiedlicher Art von der Digitalisierung profitieren. Zehn Ansatzpunkte für Digitalisierung werden im Gestaltungsbereich des Einkaufs verortet und inhaltlich skizziert. Dies soll dem Leser Gelegenheit bieten, sich im Buch an den für ihn interessanten Fragestellungen und Lösungsansätzen zu orientieren.

1.1 Gestaltungsbereich des Einkaufs

Kernaufgaben des Einkaufs sind zunächst Lieferantensuche, Verhandlung und Nominierung mit dem Ziel der Absicherung der Wettbewerbsfähigkeit für das einkaufende Unternehmen, den Abnehmer. Bei der Lieferantensuche durchsucht

F. Schupp (✉)
Bühl, Deutschland
E-Mail: schupp-florian@t-online.de

H. Wöhner
Baden-Baden, Deutschland
URL: http://www.linkedin.com/in/woehner/

© Springer Fachmedien Wiesbaden GmbH 2018
F. Schupp und H. Wöhner (Hrsg.), *Digitalisierung im Einkauf*,
https://doi.org/10.1007/978-3-658-16909-1_1

der Einkauf systematisch eine größere Grundgesamtheit an bekannten und unbekannten Lieferanten, um für einen konkreten Beschaffungsbedarf grundsätzlich
geeignete Lieferanten zu filtern. Der Suchraum sollte groß gewählt werden, um
möglichst wenig passende Lieferanten von Anfang an auszuschließen. In der
Praxis müssen je nach Anwendungsfall vierzig bis fünfzig Lieferanten durchsucht werden, um ein oder zwei passende Lieferanten zu finden. In der Folge
verhandelt der Einkauf Preis und Bedingungen. Idealerweise erfolgen derartige
Verhandlungen parallel mit mehreren Lieferanten, um eine Auswahlmöglich-
keit herzustellen. In Verbindung mit einem sich kundenseitig ergebenden Ziel-
preis wird der am besten geeignete Lieferant nominiert. Das übergeordnete Ziel
ist hierbei die Sicherstellung der Wettbewerbsfähigkeit des Abnehmers. Mit der
Nominierung stehen Lieferant und Preis fest, bezogen wird jedoch das Produkt.
Dieses wird im Wesentlichen charakterisiert durch die Qualität, die technische
Eigenschaft und Wirkung sowie die Verfügbarkeit. Damit ist es insbesondere
auch Ziel des Einkaufs, diese Aspekte zu berücksichtigen, zu entwickeln und die
gewünschte Ausprägung sicherzustellen (Abb. 1.1).

Die Qualität und technische Eigenschaft können wesentlich vor der Lieferung
beeinflusst werden. Der Einkauf stimmt dabei mit dem Lieferanten die zugrunde
liegende Qualitätsphilosophie ab, über welche der Lieferant das gewünschte Qualitätsniveau definiert. Im Zusammenspiel mit dem Abnehmer wird die Erreichung
des benötigten Qualitätsniveaus über die Qualitätsvorausplanung eingestellt.

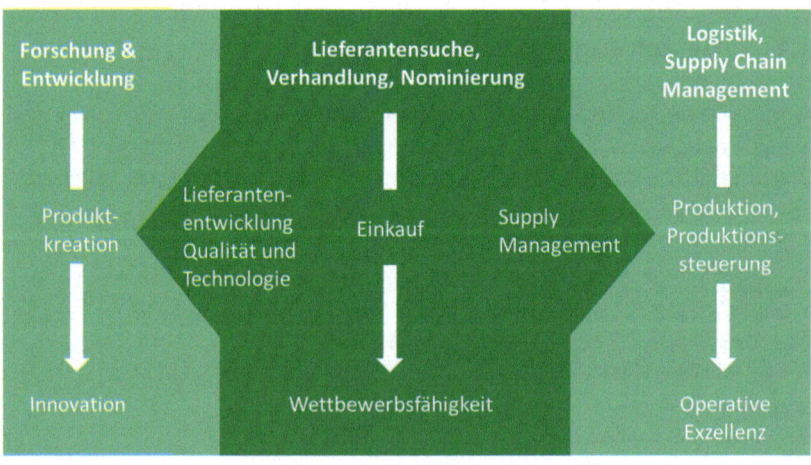

Abb. 1.1 Gestaltungsbereich des Einkaufs

Dabei ergeben sich Qualitätskosten, die aus Qualitätssicherungskosten und zu erwartenden Fehlerkosten resultieren. Die technische Produkteigenschaft wird durch Produktattribute beschrieben. Diese werden durch geeignete Fertigungstechnologien und Parameter hergestellt. Im Rahmen der Produktkreation können zusätzliche Produktattribute entwickelt werden. Da im Rahmen der abnehmerseitigen Produktkreation bestimmte Attributsausprägungen des zugekauften Produktes benötigt werden, definiert der Einkauf diese Attributsausprägungen zusammen mit dem Lieferanten. Wenn diese Attributsausprägungen nicht allein durch den Lieferanten erreicht werden können, muss der Abnehmer über den Einkauf den Lieferanten technologisch entwickeln. Umgekehrt ist der Einkauf ebenso der Entwicklungspartner für den Lieferanten, wenn der Lieferant neue Produktattribute ermöglichen kann, die zu einer neuen Produktkreation des Abnehmers führen können. In beiden Fällen ist es zwingend notwendig, dass Entwicklungsabteilung des Abnehmers und Einkauf des Abnehmers gemeinsam mit dem Lieferanten arbeiten (Schiele 2010; Tracey 2006).

Die effiziente Verfügbarkeit der lieferantenseitig hergestellten Produkte wird über das Supply Management gewährleistet. Dies hat eine qualitative Komponente und eine logistische Komponente.

Die qualitative Komponente verfolgt das Ziel, ausschließlich fehlerfreie Produkte von Lieferanten zu erhalten. Dabei wird erstens die Produktion bei Lieferanten anhand definierter Parameter überwacht, um die Anlieferung fehlerhafter Produkte zu unterbinden. Idealerweise erfolgt die Überwachung der Produktionsparameter in Echtzeit, um zunächst nicht sichtbare Fehler und Abweichungen von Attributsausprägungen auszuschließen und bei Grenzfällen vor der Anlieferung gemeinsam eine Entscheidung über die Verwendung der Produkte treffen zu können. Zweitens werden Fehler erfasst und systematisch über Ermittlung von Grundursache und Gegenmaßnahmen behoben. Hierbei überträgt das Supply Management die Erkenntnisse von Fehlern und Maßnahmen eines Lieferanten auf weitere, potenziell von ähnlichen Fehlern betroffene Lieferanten, um das Qualitätsniveau der Lieferantenbasis insgesamt nachhaltig zu heben.

In der logistischen Komponente werden physische und informatorische Warenströme mit deren Wechselwirkungen betrachtet. Produktion und Lieferungen der Lieferanten werden an der Produktion des Abnehmers ausgerichtet. Dafür werden die Zielgrößen Versorgungssicherheit, Bestandshöhe und Flexibilität abnehmerseitig definiert und produktbezogen mit dem jeweiligen Lieferanten durch jeweils passende Konzepte umgesetzt. Dabei muss zunächst ein geeignetes Kommunikationsmittel etabliert werden, das effizienten bidirektionalen Informationsaustausch ermöglicht. In einem weiteren Schritt findet die Rückkopplung aus Lieferanteninformationen Einfluss in die Produktionssteuerung des Abnehmers.

Damit kann die Produktionssteuerung des Abnehmers auf ein weiteres Optimierungsniveau gehoben werden. Das Supply Management trägt durch Transparenz und verstehbare Abläufe zur Kostenminimierung bei und ermöglicht operative Exzellenz (siehe hierzu sowie zusammenfassend zum Gestaltungsbereich des Einkaufs Abb. 1.1).

1.2   Ansatzpunkte für Digitalisierung im Gestaltungsbereich des Einkaufs

Im Folgenden werden zehn ausgewählte Ansatzpunkte für Digitalisierung innerhalb des beschriebenen Gestaltungsbereichs des Einkaufs vorgestellt. Jeder dieser Ansatzpunkte wird nachfolgend in je einem Kapitel ausführlich betrachtet.

Die folgenden Ausführungen beginnen mit der Rolle des Individuums in einem Unternehmensumfeld, das sich durch Digitalisierung wandelt. Dabei stellt die Digitalisierung für viele Unternehmen zunächst eher eine Herausforderung dar. Denn die Funktionen Einkauf, Vertrieb und Logistik – also die Unternehmensbereiche, die sich im weiteren Sinne mit Supply Chain Management befassen – sind gefordert, durch Integration die Potenziale von Big Data und Business Analytics zu nutzen. Im Kapitel „Successful digital transformations need a focus on the individual" beschreiben Prof. Alessandro Ancarani und Prof. Carmela Di Mauro von der Universität Catania, Italien, die Bedeutung von Veränderung der Unternehmenskultur, Empowerment, Mitarbeitertraining und der Notwendigkeit, in solche Führungskräfte zu investieren, die aktive Digitalisierung unterstützen.

Für kleinere und mittelgroße Unternehmen ist die aktive Teilnahme an der Digitalisierung eine große Herausforderung. Insbesondere ist es schwierig, das Wissen des Individuums immer wieder neu mit Digitalisierungswerkzeugen zu verknüpfen. Zusätzlich muss bei komplexen Fragestellungen auch das Wissen anderer Unternehmen simultan betrachtet und ausgewertet werden. An dieser Stelle beschäftigt sich das Kapitel „Boundaries of digitalization – Why companies are still using e-mail and other traditional tools to manage their knowledge – and will they continue?" aus Sicht von Lieferanten im High-Tech-Sektor mit der Frage, in welcher Weise Wissen aus der Supply Chain für Lieferanten nutzbar gemacht werden kann. Denn bei Lieferanten fehlen oftmals die notwendigen Ressourcen, um der Entwicklung der Digitalisierung zu folgen. In diesem Zusammenhang ist der Zugriff auf sogenannte Wissensmanagementsysteme beschränkt, wie Piera Centobelli, Ph.D., Prof. Roberto Cerchione und Prof. Emilio Esposito von der Universität Napoli Federico II., Italien, in einer detaillierten Untersuchung von 25 Lieferanten erörtern. Die Autoren stellen erstens fest, dass Unternehmen aus der Informations- und Kommunikationstechnologie Lieferanten nicht ausreichend unterstützen. Zweitens ergibt die Untersuchung, dass Wissensmanagement durch zum Beispiel Brainstorming, Problem Solving und Arbeitsgruppen insbesondere innerhalb der Lieferanten stattfindet, jedoch wird wenig Wissen zwischen diesen Lieferanten im Zusammenspiel mit deren Kunden generiert. Abschließend allokieren die Autoren eine Mitverantwortung für die Wissensgenerierung bei den Kunden der Lieferanten, da diese einen effizienteren Zugang zu digitalen Wissensmanagementsystemen haben.

Eine solche Verknüpfung von Lieferanten und Kunden über die gesamte Supply Chain hinweg ist notwendig, um das große Einkaufsthema „Working Capital Management" sinnvoll einer umfassenden Optimierung zuzuführen. Vorteile der Digitalisierung in diesem Zusammenhang sind vielfältig, wobei das kollaborative Handeln in der Wertschöpfungskette gebundenes Kapital freisetzen kann. In Kapitel „The effect of digitalization is measurable – How transparency on the CCC creates momentum for working capital management?" entwickeln Lotta Lind, Prof. Timo Kärri, Sari Monto, PhD, Miia Pirttilä, PhD, von der Universität Lappeenranta, Finnland, und Dr.-Ing. Florian Schupp ein Modell, um Working Capital auf Ebene der gesamten Wertschöpfungskette zu messen, zu überwachen und zu steuern. Hauptelement hierbei ist eine web-basierte Plattform, welche die physischen und finanziellen Warenströme über moderne Technologien und Echtzeitinformationen verbindet.

Die finanzielle Dimension der Digitalisierung in Unternehmensnetzwerken reicht vom erfolgreichen Working Capital Management bis zur Fragestellung, inwiefern mit dem Fortschreiten der Digitalisierung auch die potenzielle Gefahr von Unternehmensinsolvenzen oder sogar ökonomischen Krisen steigt. Dabei ist es unstrittig, dass die Digitalisierung die Gesellschaft und die Wirtschaft in den nächsten Jahren umfassend beeinflusst. In Kapitel „Digitalisierung und Krise" zeigen Dirk Adam, Wellensiek Rechtsanwälte Partnerschaftsgesellschaft mbB, Florian Glunz, PCG – Project Consult GmbH, und Prof. Klaus Kost, Ruhr-Universität Bochum und PCG – Project Consult GmbH, anhand von Thesen auf, welche perspektivischen Chancen und Herausforderungen aus ökonomischer und auch juristischer Betrachtungsweise mit der Digitalisierung verknüpft sein können. Die Digitalisierung wird zu einer Selektion und damit zu einer Verfeinerung der im Wettbewerb stehenden Unternehmen führen. Es prägen sich branchentypische Entwicklungen aus. Dabei richten die Autoren einen Appell an alle Entscheidungsträger: Die Digitalisierung muss jetzt mit allen Facetten gestaltet werden. Weitere Thesen nehmen einen vorstellbaren Krisenverlauf samt seiner Herausforderungen, Probleme und Chancen im Insolvenzfall auf. Darauf aufbauend wird das Pflichtenheft der Politik in persona eines noch nicht allseits akzeptierten „Europäischen Gesetzgebers" angesprochen.

Mit der Digitalisierung wird immer häufiger von einer neuen industriellen Revolution gesprochen, dem Zeitalter der sogenannten Industrie 4.0. Intelligente Produktionssysteme entstehen aus hoch automatisierten, selbststeuernden Maschinen und Produktionsmaterialien, welche miteinander kommunizieren. Diese Steuerungsvorgänge finden einerseits innerhalb von Unternehmen statt, andererseits treffen Objekte darüber hinaus aber auch Entscheidungen in Wertschöpfungsnetzwerken. Damit entsteht das neue Feld „Einkauf in der autonomen Produktion". Im Kapitel „Autonomous Manufacturing-related Procurement in the Era of Industry 4.0" gehen Prof. Yilmaz Uygun und Maria Ilie von der Jacobs University, Bremen, auf den Einfluss von intelligenten Produktionssystemen auf Einkaufsprozesse ein. Die Autoren unterscheiden zwischen Einkaufstätigkeiten, die vermutlich weiterhin von Menschen durchgeführt werden, und Tätigkeiten, die durch intelligente Maschinen direkt durchgeführt werden. Für Letzteres wird ein Konzept vorgelegt, welches operational voll automatisierte und produktionsorientierte Einkaufssysteme entwickelt. Neben diesen automatisierten Einkaufsprozessen bleiben innovationsbasierte Aufgaben in der Verantwortung des Einkäufers selbst.

Im Zusammenhang mit von Maschinen ausgelösten, automatisierten Bestellprozessen und unternehmensübergreifender Entscheidungsfindung spielt die digital unterstütze Absicherung von Produktqualität eine bedeutende Rolle. Die erste Herausforderung hierbei ist, die relevanten Produktionsdaten bei Lieferanten digital zu erfassen, um bezogen auf ein spezifisches Kundenprodukt Qualitätsinformationen abzuleiten und für den Kunden nutzbar zu machen. Das Kapitel „Durch Digitalisierung entsteht Qualität – Produktionsdaten bei Lieferanten für die Qualität des Kundenprodukts nutzbar machen" befasst sich mit der Digitalisierung im Umfeld von Qualitätsmanagement. Diese Idee wird von Dr. Per Larsen, DISA, Dänemark, anhand konkreter Anwendungsfälle des in der Gießereibranche tätigen Anlagenbauers DISA untersucht. Umformprozesse wie Gießen sind typischerweise mit relativ hohen Qualitätskosten verbunden, die sowohl beim Lieferanten als auch beim Abnehmer anfallen können. Solche Qualitätskosten können durch die Digitalisierung und informatorische Verknüpfung von Produktionsanlagen zur Optimierung von Produktion und damit Qualität im Gießereibetrieb reduziert werden. Dies kann durch transparente Datennutzung zwischen Gießer und Gussabnehmer sowie durch abnehmerseitigen Echtzeitzugang zu Daten der Produktionssysteme erreicht werden. In gleicher Weise kann der Abnehmer über dieselben Systeme Qualitätsdaten aus den Produktionssystemen des Abnehmers zum Lieferanten zurückspielen. Die Verwendung von Echtzeitdaten ermöglicht zeitnahe Entscheidungen, in den Produktionsprozess einzugreifen und Qualitätskosten zu vermeiden.

Neben der qualitativen Komponente profitiert auch die logistische Komponente des Supply Managements von der Digitalisierung. Dabei werden die Planungs-, Lieferungs- und Produktionsprozesse zwischen Lieferant, Spediteur und Abnehmer durch logistische Lieferantenanbindung informatorisch verzahnt. Im Kapitel „Digitalisierung in der Lieferantenanbindung" beschreibt Dr. Heiko Wöhner sieben aufeinander aufbauende Digitalisierungsstufen. In den ersten drei Stufen werden grundlegende Standards für die Digitalisierung geschaffen und bestehende Prozesse durch Digitalisierung effizienter gestaltet. Unter anderem wird die unternehmensübergreifende Sichtbarkeit der tatsächlichen Bestände sukzessive erhöht und verbesserte Steuerungsentscheidungen ermöglicht. Die folgenden zwei Digitalisierungsstufen beschreiben Verzahnungsmechanismen in Bestell- und Lieferverhalten, Transport und Produktion, die erst durch Digitalisierung ermöglicht werden. Die weiteren zwei Digitalisierungsstufen erhöhen die Effizienz des Planungssystems durch den Wandel in ein System mit dezentraler Datenhaltung und Entscheidungsfindung. In der letzten Stufe enthält die Ware selbst die digitale logistische Information.

Neben der Entwicklung von Qualität und logistischer Anbindung spielt aus Sicht des Abnehmers ebenso die technologische Entwicklung des Produktes selbst eine Rolle. In diesem Zusammenhang kann der Einkauf dabei unterstützen, zum einen die technischen Fähigkeiten des Lieferanten zu entwickeln und zum anderen Innovationen in Form gänzlich neuer Produkte oder neuer Produktattribute zu internalisieren. Hierbei unterstützt die Digitalisierung die praktikable und schnelle Messung von Wertbeiträgen aus innovativen technischen Produkteigenschaften. Damit kann die Technologiedimension über die Erfassung eines Produkteigenschaftennutzwertes neben dem Preis Einfluss auf die Vergabeentscheidung nehmen. Im Kapitel „Supplier innovation can be measured – How digitalization allows to effectively include the technology dimension into sourcing decisions" begründen Dr.-Ing. Florian Schupp und Matthias Rehm, Max Hauser und Universität Antwerpen, Belgien, warum die Produktinnovation durch Lieferanten Teil des Einkaufsgestaltungsbereichs ist. Die Autoren legen ein Modell vor, wie Nutzwerte von Produkteigenschaftskombinationen unter der Berücksichtigung von Reservationspreisen in eine Präferenzkurve des Abnehmers überführt werden können. Weiter werden die Angebote von Lieferanten in Produkteigenschaftsnutzwert-Preiskombinationen überführt und diese in einer Machbarkeitskurve dargestellt. Präferenzkurve und Machbarkeitskurve ergeben ausgerichtet am Kundenzielpreis einen Lösungsraum für Lieferanteninnovationen.

Innovationen entstehen in zunehmendem Maße außerhalb der etablierten Wertschöpfungsnetzwerke durch Unternehmen, die neu in Märkte eintreten.

Dabei werden nicht nur neue Produkte, sondern auch neue Märkte geschaffen – nicht selten werden sogar die neuen Unternehmen zu Produkten in einem übergeordneten Markt. Gerade Start-up-Unternehmen fühlen sich in diesem Umfeld wohl. Sie können mit in traditionellen Industrien unbekannten Freiheiten in kurzer Zeit Produkte entwickeln, testen und verbessern. Solche Start-ups verändern damit die Beschaffungsmärkte für Einkäufer. Die Geschäftsmodelle zahlreicher Start-ups zielen auf die direkte Verbindung von Herstellern und Konsumenten durch das eigene digitale Produkt ab. Dabei spielt die Digitalisierung die entscheidende Rolle. Im Kapitel „Opportunities in Emerging Markets – Purchasing with Start-ups" führen Dr.-Ing. Florian Schupp und Dr. Heiko Wöhner ein Interview mit Alisée de Tonnac, Seedstars. Diskutiert werden unter anderem die Fragen, wie einkaufende Unternehmen die Innovationen von Start-ups nutzen können, wie die offensichtlich bestehenden kulturellen Unterschiede einerseits genutzt und andererseits überwunden werden können, welche Rolle die Informations- und Kommunikationstechnologie für die Start-up-Produkte und die Start-up-Märkte spielt und welche enormen Potenziale es in Emerging Markets gibt. Dabei sind für Seedstars jedoch nicht die von vielen Einkäufern heute genutzten Best-Cost-Countries gemeint, sondern Länder wie Nigeria, Kenia, Peru oder Aserbaidschan.

Es ergibt sich also ein Einfluss der Digitalisierung nicht nur auf einzelne Märkte, sondern auch auf Wirtschaftsräume und Volkswirtschaften. Im abschließenden Kapitel befassen sich Philippe Gillen und Prof. Achim Wambach, Zentrums für Europäische Wirtschaftsforschung, mit „Ableitungen zum Einfluss der Digitalisierung auf die Volkswirtschaft". Als Charakteristika nennen die Autoren erstens die Dynamik, durch die sich die Digitalisierung von vorangegangenen Entwicklungen absetzt und dadurch Unternehmen fordert, einen Strukturwandel in kurzer Zeit zu bewältigen. Zweitens stehen große Mengen an allgegenwärtigen Daten im Mittelpunkt. Diese Daten sind in Verbindung mit der dritten Besonderheit, dem verstärkten Aufkommen von Plattformmärkten, zu einem Zahlungsmittel geworden. Als Reaktion hierauf wurde das Gesetz gegen Wettbewerbsbeschränkungen (GWB) novelliert. Allerdings sind die zu erwarteten Effizienzgewinne der Digitalisierung in der Produktion in den klassischen Maßen für Produktivität nicht zu sehen. Die Autoren diskutieren diesbezüglich unterschiedliche Erklärungsansätze.

Abb. 1.2 zeigt auf, auf welche Weise und an welcher Stelle die Kapitel dieses Buches in den Gestaltungsbereich des Einkaufs (s. Abb. 1.1) eingreifen.

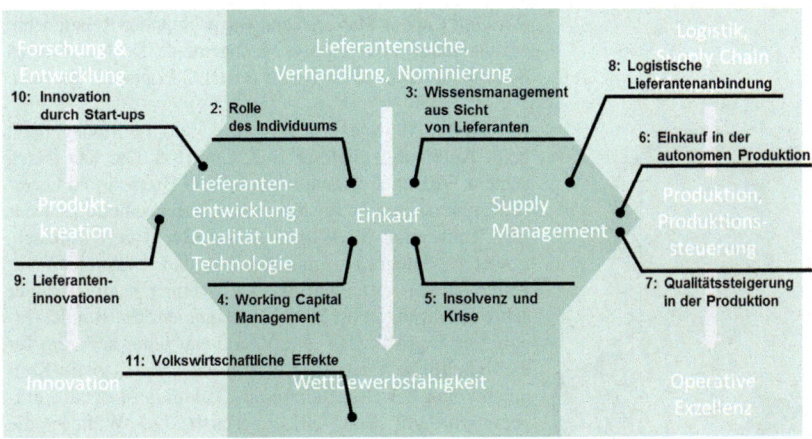

Abb. 1.2 Ansatzpunkte für Digitalisierung im Gestaltungsbereich des Einkaufs mit Kapitelnummern

Literatur

Schiele, H.: Innovationen von und mit Lieferanten – Ergebnisse einer quantitativen Studie. BME-Report, Enschede (2010)

Tracey, M.: A holistic approach to new product development: new insights. J. Supply Chain Manage. **40**(4), 37–55 (2006)

Über die Autoren

Dr.-Ing. Florian Schupp ist Einkaufsleiter Automotive und Automotive Aftermarket bei Schaeffler. In Rahmen seiner Dissertation befasste er sich an der Technischen Universität Berlin mit Themen rund um die Strategieentwicklung in Einkauf und Logistik. Dr.-Ing Florian Schupp hat insgesamt 19 Jahre Einkaufserfahrung bei Schaeffler, Continental, Siemens und SONY. Zusammen mit der Universität Lappeenranta, Finnland, und der Universität Catania, Italien, verknüpft er Einkauf und Supply Management mit wissenschaftlicher Forschung. Dr.-Ing. Florian Schupp lehrt an der Technischen Universität Berlin und der Jacobs University Bremen. Seine Forschungsschwerpunkte sind Einkaufsstrategie, Verhaltensaspekte im Einkauf, Integration von Lieferanteninnovation in Parametric-Auctions,

Working Capital Management, Buyer-Supplier-Relationship
Management und Supply Management. Dr.-Ing Florian
Schupp ist Mitglied im Beirat der BLG Logistics, Bremen.

Dr. Heiko Wöhner ist Spezialist Supply Management
beim Automobilzulieferer LuK GmbH & Co. KG. Nach
seinem Wirtschaftsingenieurstudium in Bremen und Östersund untersuchte er im Rahmen seiner Promotion an der
EBS Universität für Wirtschaft und Recht in Wiesbaden,
inwiefern Integration mit Kunden und Lieferanten für
Unternehmen vorteilhaft ist. Dr. Wöhner sammelte vier
Jahre Erfahrungen im Projektmanagement der Bundesvereinigung Logistik (BVL) e. V. und war unter anderem für
die inhaltliche Gestaltung des Deutschen Logistik-Kongresses und des Branchenforums Automobil-Logistik mitverantwortlich. Seit 2011 gestaltet Dr. Wöhner die
Lieferantenanbindung bei der LuK GmbH & Co. KG und
unterstützt Forschung im Bereich Supply Management.

Successful digital transformations need a focus on the individual

2

How does digitalization affect the behaviour of purchasers and team members in related functions?

Alessandro Ancarani und Carmela Di Mauro

Abstract

Digitalization represents a challenge for organizations. Some organizational functions such as purchasing, sales, logistics and in general all those that are part of the wider area of supply chain management will be asked to achieve a closer integration to fully exploit the potential of big data and business analytics. In order to achieve the successful implementation of digitalization, organizations need to invest in staff training, empower employees, change the organizational culture to embrace the key role of analytics for the company, and hire leaders who actively support digitalization.

2.1 Introduction

Today, manufacturing companies face a growing demand for customization of production, together with shorter lifecycles. The growing complexity of products and processes requires the development of the organization as a network that multiplies available capacities without further in-house investments, building on the fact that companies in collaborative networks can improve their agility through decoupling and spatial separation of production processes. In turn, this process of

A. Ancarani (✉) · C. Di Mauro
DICAR – Università di Catania, Viale A. Doria, 6 – 95125 Catania, Italien
E-Mail: aancaran@dica.unict.it

C. Di Mauro
E-Mail: cdimauro@unict.it

© Springer Fachmedien Wiesbaden GmbH 2018
F. Schupp und H. Wöhner (Hrsg.), *Digitalisierung im Einkauf,*
https://doi.org/10.1007/978-3-658-16909-1_2

integrating production from multiple production sources has dramatically increased the need for coordination (Jaehne et al. 2009) and for collaboration with
suppliers and customers (Ageron and Spalanzani 2010; Pilbeam et al. 2012).
Moreover, many companies maintain a competitive advantage by outsourcing
some activities to other organizations in the network while focusing on their dis-
tinctive competencies (Christopher 2000). This possibly entails offering a superior manufacturing capability and share competencies in order to exploit business
opportunities as virtual corporations, thus changing traditional business models
focused on offering superior products (Davidow and Malone 1992).

The key enabler of the changes discussed above is the widespread diffusion
of production systems that incorporate self-controlling systems using Internet and
tools for tracking individual products throughout the process chain. Within the
factory of the future (smart factory), cyber-physical systems will assist the communication of humans with machines and products. Machines able to obtain and
process data can autonomously run tasks and interact with humans via interfaces
(Brettel et al. 2014).

The collaboration between humans and machines has the potential to create
new sources of value for the business, which may determine a dynamic recon-
figuration of the company's resources. In this case, the renewal of competences
leads to a co-evolution of markets and company players (Galunic and Eisenhardt
2001). This virtuous interaction will occur if organizations, faced with the choice
to either *automate* or *informate* (Zuboff 1985), will not simply transfer tasks from
the hands of workers to machines, but rather will endow people with greater capacities that allow them to exploit the advantages of technology.

In order to increase the acceptance and actual usage of digitalization, new
organizational capabilities are required (George et al. 2016; Provost and Fawcett
2013), and leaders first have to come to recognise the complex implications of
digitalization for their company and for the employees (Wang et al. 2016). The
digital initiative places huge pressures on organizations to make major changes
in order to enhance not only individual competences but also the coordination of
persons, processes, and technologies (Desmet et al. 2015; Dörner and Meffert
2015). Therefore, the changes determined by these technologies need to be com-
plemented by changes in organizational structures, management approaches,
organizational behaviors, and operating cultures (Wade and Marchant 2014, as
reported by Kohnke 2017).

While most of the relevant technology is already available, the big issue will
be how to make possible for the organization to change according to the request
of the new business environment. In spite of the growing attention to digitalization, most contributions focus on the technology, while there is a paucity of studies

related to the "human" side of digitalization (Kohnke 2017). Discussions and analyses are mainly confined to reports and white papers by consultancy firms (among others Accenture Technology Vision 2017; Bonnet and Nandan 2011; Westerman et al. 2014).

Using the lens of organizational behaviour, this chapter aims at discussing the main challenges posed by digitalization and at identifying key issues to be addressed for successful implementation of digitalization. The main thesis is that digitalization modifies ways of working and speeds up the pace of change of the organization, requiring the rapid acquisition of new skills and competences, which in turn require building new organizational capabilities. Once organizations fulfil these requirements, they move towards digitalization by evolving their culture. The development of new leadership is a necessary pre-requisite to guide these changes. In what follows, each of these critical aspects (building competences, managing the structural change, developing new forms of leadership, and changing the organizational culture) will be discussed.

New digital technologies are expected to change all areas of the so-called traditional business processes, and in particular key business priorities like supply chain management (SCM). Expected benefits from digitalization include more informed decision-making, higher supply chain efficiency, improved demand planning, lower supply chain costs, and increased visibility (Schoenherr and Speier-Pero 2015). However, whether supply chain executives will be able to harvest the opportunities brought about by new technologies is still unclear, as most of them have not yet designed a digital strategy that "transforms their supply chains into demand-sensitive networks" (Farahani et al. 2017). Most supply chain executives agree that over the next three years supply chain performance will rely largely on their supply chain talent, which will encompass a re-design of the roles and tasks assigned to the workforce, asking the supply chain workforce to move beyond traditional roles and responsibilities. Therefore, a second aim of the chapter is to identify some of the implications of digitalization for supply chain management (SCM) and functions related to SCM (logistics, operations management, purchasing).

The chapter is organized into three main sections. The first introduces an appropriate framework for the analysis of the interplay technology-human resources as a source of competitive advantage and identifies some of the new capabilities required. The second discusses the issue of managing the necessary organizational changes, and the final one highlights some of the key actions the leader of the digital era should undertake to achieve profound and long-lasting results.

2.2   Developing dynamic capabilities

In the past ten years, digital technologies have increasingly been incorporated by organizations, driven by advancements in technology. Organizations are struggling to keep up with the accelerating pace of change determined by digitalization (McAfee and Brynjolfsson 2012; Fitzgerald et al. 2014; Kotter 2014), which has radically changed how employees work, communicate, and collaborate, and has created a challenge for leadership (Colbert et al. 2016).

Although digital tools have been inside companies already for many years, what has changed recently is the acceleration of both the capabilities enabled by these tools as well as the pace of adoption by customers, employees, and organizations alike, which in turn are having a profound impact on organizations. The organizational processes and hierarchical structures that companies have used in the past are no longer effective, since they can actually hinder attempts to compete in a marketplace where discontinuities are more frequent and companies are required constantly to adjust to changing contexts (Kotter 2012).

The increasing turbulence and uncertainty of the external environment ask for re-definition of the competitive advantage of firms by exploring new opportunities. "Dynamic capabilities" are called to renovate competences in order to achieve alignment with the changing environment (Teece 2007; March 1991). Dynamic capabilities can be interpreted as the high-order capabilities that include several different capacities, such as the capacity for improving quality, the capacity for managing human resources, and the capacity for utilizing technologies (Drnevich and Kriauciunas 2011; Helfat and Peteraf 2009). According to Teece, dynamic capabilities, a concept rooted in the resource-based view of competitive advantage (Barney 1991; Newbert 2007), explain why some firms outperform others and experience continuous competitive advantage (Teece 2007). Within the dynamic capabilities framework, sustainable advantage requires unique and difficult-to-replicate capabilities, including those required to dynamically adapt to changing customers and technological opportunities and embracing the company's capacity to shape the ecosystem it occupies, develop new products and processes, and design and implement viable business models.

The Information Technology (IT) literature has investigated how IT can be leveraged for the creation of dynamic capabilities, leading to increases in operational performance (Pavlou and El Sawy 2011). The extensive application of digital tools has led to the emergence of big data business analytics (BDBA), as a critical capability to provide organizations with enhanced resources to obtain value and gain a competitive advantage (Chen et al. 2012). Drawing from the concept of dynamic capabilities, BDBA can be understood as an IT-enabled, analytical dynamic capability for improving firm performance. It incorporates two dimensions: big data (BD) and business analytics (BA). Big data can be defined as a holistic approach focused on managing, processing and analysing data in order to create actionable insights for sustained value delivery, measuring performance, and establishing competitive advantages. Big data refers to the capability to process data characterized by the five Vs (volume, velocity, variety, veracity, value). As a matter of fact, big data are associated to an enormous *volume* of data, needing a significant *velocity* of processing to be effective, coming from a *variety* of sources, which need to be verified for *veracity*, and whose *value* has to be carefully extracted.

Business analytics enables realization of business objectives identifying trends, creating predictive models for forecasting and optimizing business processes (Accenture Global Operations Megatrends Study 2014; Mithas et al. 2011; Robinson et al. 2010; Singh 2003).

In the area of purchasing, BDBA has been used to support decision-making especially in logistics and supply chain management (Muhtaroglu et al. 2013; Waller and Fawcett 2013). Its goals are to improve the flexibility, visibility, and integration of global supply chains and logistics processes, by improving the management of demand and cost volatility (Columbus 2015). In purchasing and supply chain management, where a huge amount of data is continuously generated from internal or external sources (Auramo et al. 2005), BDBA offers the opportunity to summarise and integrate this information. With particular reference to supply chain management, data management is the pillar of what is called supply chain analytics (SCA) oriented to improve supply chain operations and performance by means of database and analytical technologies (Fernandes et al. 2013, Guerrero et al. 2013; Gustavsson and Wanstrom 2009). Supply chain coordination practices further add to the data to be tracked (Souza 2014; Hayya et al. 2006; Jonsson and Mattsson 2008; Yeniyurt et al. 2013). Therefore, companies need to have the capability to store, manage and transform data and to provide access to the data for performance management activities and for supply chain operations (Chen et al. 2006; Chae and Olson 2013; Oruezabala and Rico 2012). This data management capability is particularly relevant in uncertain business environments (Davis-Sramek et al. 2010; O'Dwyer and Renner 2011) as it enables other capabilities in analytical process and supply chain performance management (Azvine et al. 2005; Sahay and Ranjan 2008). The analytical process relies on the representation of the supply chain as a series of processes (plan, source, make, and deliver), which requires companies to possess the analytical capability to handle each of them (Trkman et al. 2010). Finally, high performing supply chains require performance measurement capabilities in order to be effective in observing, orienting, and deciding on how to perform company's tasks (Chae and Olson 2013).

Until recently, the typical automated operations systems (e.g. Enterprise Resourcing Planning) could not complete a whole process, therefore requiring *knowledge workers* to extract large sets from one system and move them to another, or to intervene manually on the system in ways that may be at odds with the system logic. "Robotic Process Automation" promises to add value by performing the kinds of administrative tasks that normally require human handling, such as transferring data from multiple input sources (e.g. email and spreadsheets) to operational systems, therefore liberating time from knowledge workers for higher order tasks (Lacity and Willcocks 2015). What will be required for the decision maker is to choose which sub-systems he/she wants to coordinate and in which respect. The knowledge that is required is not necessarily related to computing or programming but rather the understanding of the company's goals, operations, and markets (see software like Blue Prism). Therefore, the best workers will not be those who can digest figures and statistics but those who have an inquisitive mind-set, can develop insights from numbers, and exercise critical thinking and act.

As the costs of Robotic Process Automation and business analytics fall, company work will be organised more and more in teams of human-robot with a specialised division of tasks, where menial, data gathering and summarising tasks will be undertaken by machines, and creative tasks and data interpretation and use should be left to humans. However, a necessary step of this process envisages managers and workers viewing robots not as competitors but as collaborators in problem solving and value creation.

2.3 Managing the change

According to business analysts and academic studies alike, one of the main shortcomings of current digital transformation programs is that they are focused on technology, targeting the automation of existing business processes rather than trying to reengineer them to generate new business models (Bonnet and Nandan 2011). Further, organizations miscalculate the dynamics of digitalization, which require the continuous alignment of people, technologies, organizational structures, and eventually organizational culture. They fail in recognizing that the critical obstacle for digital transformation is people, not technology (Bonnet and Nandan 2011). In this respect, transformations concentrated on technology, and separating technology development and implementation from the everyday processes and practices of business functions, end up giving rise to a substantial internal opposition to change.

Digital transformations entail complex challenges associated with the overall competitive strategy of the company, with the changes in the organizational system (structures, processes, and culture), and with the way changes must be managed.

The digital strategy needs a new *vision* and a capability to drive the change throughout the organization (Bonnet and Nandan 2011; Westerman et al. 2014) and to translate the vision into operational and tangible objectives. Managers should provide this vision, so that the use of BDBA becomes embedded in the way the business operates, and thus greater business analytics maturity can be achieved through a reinforcement process between the improvement of company's performance and the increasing capability of data analysis (Trkman et al. 2010). The new vision should also support employees understanding whether and how to change their behaviour.

Under volatile, uncertain, complex, and ambiguous (VUCA) conditions, like the ones faced by companies in many sectors, digitalization permits and forces leaders to make continuous adjustments in their strategy and in their business models (Horney et al. 2010). Organizational structure and processes must adapt to suit these changes. Digital transformations are inherently cross functional and rapid, and they are hindered by a hierarchical structure of organizations and a "silo thinking" style decision-making (McAfee and Welch 2013). In fact, digitalization raises the level of transparency through the free-flow of information, thus making managers perceive a loss of control, especially at mid-level, and a threat to their leadership role, eventually determining a resistance against digitalization. Therefore, an organizational structure focused on supporting digitalization must be collaborative, flexible, and agile to a sufficient degree, while keeping the rest of the activities running efficiently (Desmet et al. 2015). Organizations need a capability to be at the same time dynamic/ adaptive and stable (resilient, reliable, and efficient), combining "hierarchy and flexible network to form a twofold operating system" (Kotter 2014).

As digitalization is speeding up the pace of change and modifying the way of working, it is also important to adapt to digitalization the way of delivering *change management* (McAfee and Welch 2013). A first step in this direction is the creation and preservation of *a sense of urgency* around digital programs that engages people in activating the dynamic capabilities of the organization in order to reconfigure the existing operational capabilities and deploy new ones (Kotter 2007).

Another important step for promoting digitalization organization-wide is the development of a *communication plan*. It is fundamental to identify those employees willing to help because they are enthusiastic and open to digitalization (Fitzgerald et al. 2014). These "digital champions" can facilitate the connection between the digital transformation programs supported by the top management and the various business functions and departments. Further, they can help in transferring the knowledge to other employees (McAfee and Welch 2013). In this respect, the support of digital tools for connecting with employees in new ways (e.g. social networks) may be a simple way of embracing all employees in the change. In the same vein, digital champions and the use of social networks can be suitable ways to get suppliers involved in the digitalization process by easing knowledge sharing between buyer and supplier and creating a collaborative and dynamic environment (Trimi and Galanxhi 2014).

Two recent surveys by consultancy companies (Bughin et al. 2015; Fitzgerald et al. 2014) have confirmed the above agenda, highlighting that the main shortcoming in the digital transformation is the lack of urgency in managing the change. Next follows lack of vision, inability of organizations to keep pace with the speed of digitalization, failure to embrace a mind-set prone to experimentation, organizational culture that is not open to change uncertain roles and responsibilities, and shortage of leadership skills.

Another important step in sustaining the digitalization process consists in *aligning the systems of incentives* and key performance indicators (KPIs). People need to be activated and engaged adapting the systems of incentives and rewards to support organizations in developing a high-performance culture and in translating the digital vision into a set of measures and objectives to monitor the advancement towards digitalization (Bughin et al. 2015). Researchers have most typically assessed the adoption and use of digital services focusing on the characteristics those digital services possess, such as perceived ease of use and perceived usefulness (Ancarani et al. 2010; Venkatesh and Hala 2008). Most studies confirm that the positive effects of IT have been explicitly and positively correlated to its usage (Trice and Treacy 1988). In particular, some researchers have identified usage as the main antecedent of the impact of IT on performance (Devaraj and Kohli 2003). However, people are often unwilling to change and to adopt IT, even if it could benefit their job performance, because they do not have sufficient knowledge of the technology or do not receive *adequate support* from the work environment. In order to foster usage, KPIs should measure actual user behaviour, in terms of frequency and intensity, rather than availability of digital tools.

Finally, even though companies commonly instruct and sometimes even mandate the way professionals use technology in organizational settings, the worker's personal orientation toward technology impacts on the way professionals use technology to carry out work related tasks (Hallikainen et al. 2017). In line with this view, the survey undertaken in the US by Michigan State University on SCM professionals suggests that SCM professional's personal beliefs about the value of SCM predictive analytics is one of the primary drivers for early adoption (Schoenherr and Speier-Pero 2015).

2.4   New challenges for leaders

Aligning leadership on digitalization is a necessary basis for organizational and strategic change, as top management commitment represents a critical success factor. We identify two key actions the digital leaders should undertake – first: empower employees, next: support a shift of the organizational culture.

Employee empowerment
Effectiveness in BDBA should include developing high-level skills (Schoenherr and Speier-Pero 2015), which allow the use of the new generation of IT tools and architectures to collect data from various sources. Data should be stored, organized, extracted, and analysed in order to generate valuable insights by sharing them with stakeholders for competitive advantage co-creation. Therefore, the development of new digital business models calls for the empowerment of the employees to leverage digital technologies and applications (Westerman et al. 2012).

The skills must encompass both domain knowledge (e.g. purchasing, marketing) and analytic knowledge, which must be acknowledged as both relevant and indivisible. The data scientist needs both: a deep domain knowledge and a broad set of analytical skills. Although there is not a strong body of evidence yet, findings from a survey carried out among experts in different manufacturing industries suggest that there are decreasing returns on domain knowledge, even if domain knowledge is still a pre-requisite for data analysis. Further, the width of the analytical skills moderates the relationship between the domain knowledge of data scientists and their effectiveness (Waller and Fawcett 2013).

Employees should also be provided with essential competencies for dealing with the complexity, amplified speed, and intensity of change that digitalization determines. Therefore, while in the future organizations might be willing and able to look to the open market to employ the right persons, internal investments in training and education in digital technologies remain a crucial success factor for any transformation program (Hoberg et al. 2015). For example, in Italy, according to the survey conducted by Randstad Workmonitor (2016), while 90% of respondents agree on the fact that companies should develop a digital strategy, they suggest that only 30% of them have employees with the required skills, and 67% define as scarce their own level of digital competencies.

Finally, one interesting avenue through which employee empowerment can be realized is through talent management. Pioneering companies are already enhancing their competitive advantage by adopting methods to analyse employee talent. For example, the combination technology/humans is expected to enable humans to become "intra-preneurs" (internal entrepreneurs) or dynamic innovators who can proactively identify growth opportunities for their company and manage the associated risks. In this respect, supply chain management and supply planning constitute critical areas for intra-preneurship, determining an expected increase in the number of jobs available in supply chain management in areas such as demand-supply balancing, category/segment management, supply chain risk management, and supply chain analysis (Accenture 2017). Most significant shifts envisaged in roles encompass: i) planners spending more time on forward-looking strategic decisions and less on reactive problem solving; ii) logistics managers shifting to be supply chain managers and working to integrate related functions. More in general, businesses are taking steps to design a new role for workers in the digital era, as companies embrace the wide spectrum of worker-relationship types in the open talent marketplace, from independent contractors, to full-time employees, to every variation in between.

Shift of culture

Digital programs require new forms of leadership able to unfreeze the existing cultures and make them evolve towards cultures oriented to technology driven innovation. Therefore, as for most strategic changes, the success of digital initiatives depends on organizations leaders' commitment. Leaders that value human-capital insights should foster an open culture valuing experimentation and not sanctioning mistakes (Schein 2010), which is at odds with how many human resource functions work currently. An organizational culture that encourages risk taking is required for strong digital performance, and, in this direction, leadership is crucial in starting to create this culture (Schein 2010). Further, and consistent with the above discussion on talent management, the turnaround in human resource management should look at employees also as a source of collective data (e.g. work climate and employee satisfaction, productivity, retention, etc.) that managers can use to make better decisions about their capabilities and motivations (George et al. 2016). Data-driven insights on the human resources satisfaction and talent promise to deliver inimitable competitive advantage (Davenport 2009; Davenport et al. 2010).

The impact of big data on how decisions are taken and on who gets to take them is another critical aspect of the cultural change brought about by digitalization. Traditionally, when data are scarce, expensive to obtain, or not available in digital form, the role of experts (including high ranking officers in the organization) is to make decisions with the so called "intuitive" approach, based on the experience they have built and on the patterns and relationships they have observed and internalized. Domain expertise remains a critical asset when it comes to know where the biggest opportunities and challenges lie. However, as big data are progressively available, the role of domain experts will shift from using the intuitive approach to the definition of what questions should be asked to the data in order to identify opportunities and threats (McAfee and Brynjolfsson 2013). This will require a profound shift of leadership, away from top management intuition and towards data-driven bottom-up decision making.

2.5 Conclusions

Digitalization has changed work, communication, and collaboration modes in the workplace. Therefore, it represents a challenge for all organizations, requiring an adaptation of structures, strategies, leadership and culture. Some organizational functions such as purchasing, sales, logistics, and in general all those that are part of the wider area of supply chain management will be asked to achieve a closer integration. In fact, one of the key advantages of BDBA is the possibility to interconnect data from different sources and different sub-systems.

The relatively low percentage of SCM specialists actively using supply chain analytics (Schoenherr and Speier-Pero 2015) is a testament to the need for many companies to undertake a three-step path implicitly suggested in this chapter. To summarize, the three key steps consist of:

a. investing in training (not only technologies) in order to effectively empower employees with new capabilities leading to the effective usage of digital tools;
b. working on individual attitudes by changing the organizational culture so to embrace the key role of analytics for the company;
c. choosing leaders who actively support digitalization and the use of analytics in their decisions.

A closer collaboration between educational institutions such as universities and business may contribute to fill the gap between present competencies and those required for a full exploitation of the potential of digitalization.

References

Accenture Global Operations Megatrends Study Big Data Analytics in Supply Chain: Hype or Here to Stay? http://www.accenture.com/us-en/Pages/insight-global-operations-megatrends-big-data-analytics.aspx (2014). Accessed 14. Febr. 2017

Accenture Technology Vision 2017 Technology for People, AmplifYou. https://www.accenture.com/us-en/insight-disruptive-technology-trends-2017 (2017). Accessed 14. Febr. 2017

Ageron, B., Spalanzani, A.: Value creation and supplier selection: An empirical analysis. In: Wang, L., Koh, S.C.L. (Hrsg.) Enterprise Networks and Logistics for Agile Manufacturing, S. 137–153. Springer, London (2010)

Ancarani, A., Di Mauro, C., Giammanco, M.D., Mascali, F.: The use of information technology in health organisations procurement. 4th International Public Procurement Conference (IPPC2010) (1–19) (2010)

Auramo, J., Kauremaa, J., Tanskanen, K.: Benefits of IT in supply chain management: an explorative study of progressive companies. Int. J. Phys. Distrib. Logistics Manage 35(2), 82–100 (2005)

Azvine, B., Nauck, D., Cui, Z.: Towards real-time business intelligence. BT Technol. J 23(3), 214–225 (2005)

Barney, J.: Firm resources and sustained competitive advantage. J Manage 17(1), 99–120 (1991)

Bonnet, D., Nandan, P.: Transform to the power of digital – digital transformation as a driver of corporate performance. https://www.capgemini.com/resources/transform-to-the-power-of-digital (2017). Accessed 21. Febr. 2017

Brettel, M., Friederichsen, N., Keller, M., Rosenberg, M.: How virtualization, decentralization and network building change the manufacturing landscape: An Industry 4.0 Perspective. International Journal of Mechanical, Industrial Science and Engineering 8(1), 37–44 (2014)

Bughin J., Holley, A., Mellbye, A.: Cracking the digital code: McKinsey global survey results. Available via McKinsey & Company. http://www.mckinsey.com/business-functions/business-technology/our-insights/cracking-the-digital-code (2015). Accessed 21. Febr. 2017

Chae, B., Olson, D.L.: Business analytics for supply chain: A dynamic-capabilities framework. Int. J. Inf. Technol. Decis. Making 12(01), 9–26 (2013)

Chen, L., Long, J., Yan, T.: E-supply chain implementation strategies in a transitional economy. Int. J. Inf. Technol. Decis. Making 5(2), 277–295 (2006)

Chen, H., Chiang, R.H., Storey, V.C.: Business intelligence and analytics: From big data to big impact. MIS Q. 36(4), 1165–1188 (2012)

Christopher, M.: The agile supply chain. Ind. Mark. Manage. 29(1), 37–44 (2000)

Colbert, A., Yee, N., George, G.: The digital workforce and the workplace of the future. Acad. Manage. J **59**(3), 731–739 (2016)

Columbus, L.: Ten ways big data is revolutionizing supply chain management. Forbes (13. July 2015)

Davenport, T.: Make better decisions. Harv. Bus. Rev. **87**(11), 117–123 (2009)

Davenport, T.H., Harris, J., Shapiro, J.: Competing on talent analytics. Harv. Bus. Rev. **88**(10), 52–58 (2010)

Davidow, W., Malone, M.: The virtual corporation. Harper Collins, New York (1992)

Davis-Sramek, B., Germain, R., Iyer, K.: Supply chain technology: the role of environment in predicting performance. J. Acad. Mark. Sci. **38**, 42–55 (2010)

Desmet, D., Duncan, E., Scanlan, J., Singer, M.: Six building blocks for creating a high-performing digital enterprise. Available via McKinsey & Company. http://www.mckinsey.com/businessfunctions/organization/our-insights/six-building-blocks-for-creating-a-high-performing-digitalenterprise (2015). Accessed 19. Febr. 2017

Devaraj, S., Kohli, R.: Performance impacts of information technology: is actual usage the missing link? Manage. Sci. **49**(3), 273–289 (2003)

Dörner, K., Meffert, J.: Nine questions to help you get your digital transformation right. Available via McKinsey & Company. http://www.mckinsey.com/business-functions/organization/our-insights/nine-questions-to-help-you-get-your-digital-transformation-right (2015). Accessed 19. Febr. 2017

Drnevich, P., Kriauciunas, A.: Clarifying the conditions and limits of the contributions of ordinary and dynamic capabilities to relative firm performance. Strateg. Manage. J **32**, 254–279 (2011)

Farahani, P., Meier, C., Wilke, J.: Digital supply chain management agenda for the automotive supplier industry. In: Oswald, G., Kleinemeier, M. (Hrsg.) Shaping the digital enterprise, pp. 157–172. Springer International Publishing, Switzerland (2017)

Fernandes, R., Gouveia, B., Pinho, C.: Integrated inventory valuation in multi-echelon production/distribution systems. Int. J. Prod. Res. **51**(9), 2578–2592 (2013)

Fitzgerald, M., Kruschwitz, N., Bonnet, D., Welch, M.: Embracing digital technology: a new strategic imperative. MIT Sloan Manage. Rev. **55**(2), 1 (2014)

Galunic, D.C., Eisenhardt, K.M.: Architectural innovation and modular corporate forms. Acad. Manage. J **44**(6), 1229–1249 (2001)

George, G., Osinga, E., Lavie, D., Scott, B.: Big data and data science methods for management research. Acad. Manage. J **59**(5), 1493–1507 (2016)

Guerrero, W.J., Yeung, T.G., Guret, C.: Joint-optimization of inventory policies on a multi-product multi-echelon pharmaceutical system with batching and ordering constraints. Eur. J. Oper. Res. **231**(1), 98–108 (2013)

Gustavsson, M., Wanstrom, C.: Assessing information quality in manufacturing planning and control processes. Int. J. Qual. Reliab. Manage. **26**(4), 325–340 (2009)

Hallikainen, H., Paesbrugghe, B., Laukkanen, T., Rangarajan, D., Gabrielsson, M.: How individual technology propensities and organizational culture influence B2B customer's behavioural intention to use digital services at work? Proceedings of the 50th Hawaii International Conference on System Sciences (2017)

Hayya, J., Kim, J., Disney, S., Harrison, T., Chatfield, D.: Estimation in supply chain inventory management. Int. J. Produ. Res. **44**(7), 1313–1330 (2006)

Helfat, C., Peteraf, M.: Understanding dynamic capabilities: progress along a developmental path. Strat. Organiz. **7**(1), 91–102 (2009)

Hoberg, P., Krcmar, H., Oswald, G., Welz, B.: Skills for digital transformation – research report. Initiative for digital transformation (IDT) at the Technical University of Munich, Chair for Information Systems (2015)

Horney, N., Pasmore, B., O'Shea, T.: Leadership agility: a business imperative for a VUCA world. People Strat **33**(4), 32–38 (2010)

Jaehne, D.M., Li, M., Riedel, R., Mueller, E.: Configuring and operating global production networks. Int. J. Produ. Res. **47**(8), 2013–2030 (2009)

Jonsson, P., Mattsson, S.A.: Inventory management practices and their implications on perceived planning performance. Int. J. Produ. Res. **46**(7), 1787–1812 (2008)

Kohnke, O.: It's not just about technology: the people side of digitization. In: Oswald, G., Kleinemeier, M. (Hrsg.) Shaping the digital enterprise, pp. 69–91. Springer International Publishing, Switzerland (2017)

Kotter, J.P.: Leading change – why transformation efforts fail. Harv. Bus. Rev. **85**(1), 96 (2007)

Kotter, J.P.: Change faster. Harv. Bus. Rev. **2012**(11), 45–58 (2012)

Kotter, J.P.: Accelerate – building strategic agility for a faster-moving world. Harvard Business Review Press, Boston (2014)

Lacity, M., Willcocks, L.: What knowledge workers stand to gain from automation. Harv. Bus. Rev. 19. June 2015

March, J.G.: Exploration and exploitation in organizational learning. Organ. Sci **2**(1), 71–87 (1991)

McAfee, A., Brynjolfsson, E.: Big data: the management revolution. Harvard Business Review **2012**(10), 59–68 (2012)

McAfee, A., Welch, M.: Being digital: engaging the organization to accelerate digital transformation. Digit. Transform. Rev. **4**, 37–47 (2013)

Mithas, S., Ramasubbu, N., Sambamurthy, V.: How information management capability influences company performance. MIS Q. **35**(1), 237–256 (2011)

Muhtaroglu, F.C.P., Demir, S., Obali, M., Girgin, C.: Business model canvas perspective on big data applications. Big Data, 2013 IEEE International Conference on, pp. 32–37. IEEE (2013)

Newbert, S.: Empirical research on the resource-based view of the firm: An assessment and suggest ions for future research. Strat. Manage. J **28**, 121–146 (2007)

O'Dwyer, J., Renner, R.: The promise of advanced supply chain analytics. Supply Chain Management Review **15**, 32–37 (2011)

Oruezabala, G., Rico, J.C.: The impact of sustainable public procurement on supplier management: the case of French public hospitals. Industr. Mark. Manage. **41**(4), 573–578 (2012)

Pavlou, P.A., Sawy, O.A.El: Understanding the elusive black box of dynamic capabilities. Decis. Sci. **42**(1), 239–273 (2011)

Pilbeam, C., Alvarez, G., Wilson, H.: The governance of supply networks: a systematic literature review. Supply Chain Management: An International Journal **17**(4), 358–376 (2012)

Provost, F., Fawcett, T.: Data science and its relationship to big data and data-driven decision making. Big Data **1**(1), 51–59 (2013)

Randstat Global Report Randstat Workmonitor – Q6 2016. Digital awareness is rising, but we're not there yet. Group Communication Randstat Holding nv. (2016)

Robinson, A., Levis, J., Bennett, G.: INFORMS to officially join analytics movement. OR/MS Today **37**(5), 59 (2010)

Sahay, B.S., Ranjan, J.: Real time business intelligence in supply chain analytics. Inform. Manage. Comput. Secur. **16**(1), 28–48 (2008)

Schein, E.H.: Organizational culture and leadership, 4. Aufl. Jossey-Bass, San Francisco (2010)

Schoenherr, T., Speier-Pero, C.: Data science, predictive analytics, and big data in supply chain management: current state and future potential. J. Bus. Logistics **36**(1), 120–132 (2015)

Singh, N.: Emerging technologies to support supply chain management. Communications of the ACM **46**(9), 243–247 (2003)

Souza, G.C.: Supply chain analytics. Business Horizons **57**(5), 595–605 (2014)

Teece, D.: Explicating dynamic capabilities: The nature and microfoundations of (sustainable) enterprise performance. Strat. Manage. J **28**, 1319–1350 (2007)

Trice, A.W., Treacy, M.E.: Utilization as a dependent variable in MIS research. ACM SIG-MIS Database **19**(3–4), 33–41 (1988)

Trkman, P., McCormack, K., Oliveira, M.de, Ladeira, M.: The impact of business analytics on supply chain performance. Decis. Support Syst. **49**(3), 318–327 (2010)

Trimi, S., Galanxhi, H.: The impact of Enterprise 2.0 in organizations. Serv. Bus. **8**(3), 405–424 (2014)

Venkatesh, V., Bala, H.: Technology acceptance model 3 and a research agenda on interventions. Decis. Sci. **39**(2), 273–315 (2008)

Wade, M., Marchant, D.: Are you prepared for your digital transformation? Understanding the power of technology AMPS in organizational change. Tomorrow's challenges. IMD Lausanne, Switzerland (2014)

Waller, M.A., Fawcett, S.E.: Data science, predictive analytics, and big data: A revolution that will transform supply chain design and management. J. Bus. Logistics **34**(2), 77–84 (2013)

Wang, G., Gunasekaran, A., Ngai, E.W., Papadopoulos, T.: Big data analytics in logistics and supply chain management: Certain investigations for research and applications. Int. J. Produ. Econ. **176**, 98–110 (2016)

Westerman, G., Bonnet, D., McAfee, A.: Leading digital: turning technology into business transformation. Harv. Bus. Rev. Press, Boston (2014)

Westerman, G., Tannou, M., Bonnet, D., Ferraris, P., McAfee, A.: The digital advantage: how digital leaders outperform their peers in every industry. Capgemini Consulting and the MIT Center for Digital Business Global Research, London (2012)

Yeniyurt, S., Henke, J.W., Cavusgil, E.: Integrating global and local procurement for superior supplier working relations. Int. Bus. Rev. **22**(2), 351–362 (2013)

Zuboff, S.: Automate/Informate: the two faces of intelligent technology. Organ. Dyn. **14**(2), 5–18 (1985)

About the Authors

Alessandro Ancarani, Ph.D., is Associate Professor in Managerial Engineering, Università di Catania, Italy. He was co-editor in chief of the Journal of Purchasing and Supply Management (2010–2015). He is member of the Board of the Journal of Public Procurement and of the European Journal of Public Procurement Markets. He was President of IPSERA in 2009–2012 and founder member of EDSI. His research interests are in the analysis of intangibles in public service organizations, public procurement, suppliers' performance evaluation, behavioral operations and manufacturing location decisions. He has published in leading journals such as British Journal of Management, International Journal of Production Economics, International Journal of Production & Operations Management, and Social Science & Medicine.

Carmela Di Mauro, D.Phil., is Associate Professor in Business and Management Engineering, University of Catania, Italy. Her research focuses on organizational behavior and behavioral operations, global sourcing and reshoring, and health care organization and management. Her most recent research has been published in international referred journals, including The International journal of production economics, British journal of management, Journal of purchasing and supply management, Technological forecasting and social change. She is an associate editor of the Journal of Purchasing and Supply Management. She is member of the executive board and immediate past-president of the European division of the Decision Sciences Institute.

Boundaries of digitalization – Why companies are still using e-mail and other traditional tools to manage their knowledge – and will they continue?

3

Piera Centobelli, Roberto Cerchione und Emilio Esposito

Abstract

This chapter identifies relevant digital tools and managerial practices supporting firms in the different phases of knowledge management process. Based on a survey that involved 25 suppliers operating in high-tech industries, this chapter highlights that firms are generally inclined to use not updated knowledge management systems (KMS) instead of the newer ones, which are also cheaper and user friendly. This gap shows the difficulties that suppliers have to be responsive to the rapid technological changes as well as the lack of support from information and communication technology (ICT) vendors in the decision making process for choosing adequate IT-based tools (KM-Tools). The chapter also highlights that the majority of suppliers investigated adopts a variety of knowledge management practices (KM-Practices) that are used on average more intensely than KM-Tools. The field analysis demonstrates that many practices are common to many suppliers and that customers play a crucial role in supporting suppliers in their knowledge management process.

P. Centobelli · R. Cerchione (✉) · E. Esposito
Napoli, Italien
E-Mail: roberto.cerchione@uniparthenope.it

P. Centobelli
E-Mail: piera.centobelli@unina.it

E. Esposito
E-Mail: emilio.esposito@unina.it

© Springer Fachmedien Wiesbaden GmbH 2018
F. Schupp und H. Wöhner (Hrsg.), *Digitalisierung im Einkauf*,
https://doi.org/10.1007/978-3-658-16909-1_3

3.1 Introduction

Knowledge management (KM) is a crossroad research area that involves a variety of different subject areas, such as Business and Economics, Computer Science, Decision Sciences, Engineering. Despite that, there is an increasing literature analyzing how knowledge management is becoming a key strategic factor in the new industrial environment. In the field of supply management the role of knowledge management still seems to be neglected. Whereas a consistent amount of research concerning knowledge management in large customer companies exists, the literature has devoted little attention to supply firms. In particular, while the literature proposes a variety of contributions underlining the critical success factors affecting KM (Desouza et al. 2003; Blome et al. 2014; Chang et al. 2012; Cheng and Fu 2013; He et al. 2013; Kim et al. 2012; Loke et al. 2012; Mak and Ramaprasad 2003; Zhang and Zhou 2013) and the impact of KM on performance (Abid and Ali 2014; Fugate et al. 2012; Halley et al. 2010; Hult et al. 2004, 2006; Liu et al. 2012; Lu et al. 2014; Sambasivan et al. 2009), there is little empirical research on knowledge management systems (KMS) used by suppliers to support the knowledge management process. The few papers that deal with the topic of KMS in supply chain do not offer a clear framework about the KMS used and do not highlight their degree of diffusion and their intensity of use. This aspect is crucial, since nowadays technology dynamics and innovation in information and communication technology (ICT) are driving the development and the introduction of new KMS that are cheaper, easier to use and more effective than traditional ones (Garrigos-Simon et al. 2012).

In this context, the aim of this research is to investigate their individual degree of diffusion and the intensity of use of KMS through a survey based on a sample of selected suppliers.

The chapter is structured in six sections. After the introduction, in the second section the literature review is illustrated. The methodology is analyzed in the third section. The results are described in the fourth section. Finally, conclusions and implications are presented.

3.2 Literature review

In the literature there is not a common definition of KMS, but there is a variety of definitions that ranges between two extremes: (1) IT-based systems to support methods and techniques of KM (Alavi and Leidner 2001); (2) an information system or a managerial practice adopted to support companies in creating, storing or transferring knowledge, with the purpose to improve efficiency and effectiveness of knowledge management processes (Corso et al. 2003).

Summarising these two contributions and according to a KMS includes both the IT-based tools and the organisation practices. Specifically, in our approach KMS are divided into two parts: knowledge management practices (KM-Practices), defined as the set of methods and techniques to support the organisational processes of knowledge creation, storage, transfer; and knowledge management tools (KM-Tools), namely the specific IT-based systems supporting KM methods and techniques. This definition has the quality to be more comprehensive and includes both the IT-based tools and the organisation practices identified in the two previous definitions.

To illustrate the state-of-the-art on knowledge management systems in supply firms and according to the systematic literature review on KM in supply chain conducted by the literature on the topic was studied in detail from two perspectives:

- process perspective;
- unit of analysis perspective.

As for the *process perspective*, according with Kanat and Atilgan (2014) the process of knowledge management has been subdivided into 3 phases: (1) knowledge creation, in which knowledge is acquired and validated; (2) knowledge storage, in which knowledge is retained and organised; and (3) knowledge transfer, in which several actors exchange and share knowledge.

Regarding the *unit of analysis perspective*, KMS have been divided into two parts: knowledge management tools (KM-Tools) and knowledge management practices (KM-Practices), which are respectively the technical and the organisational support to knowledge management development (Alavi and Leidner 2001; Corso et al. 2003).

As for the first perspective, most of the papers that analyse the topic of knowledge management systems in supply firms focus on knowledge transfer phase. With regard to the second perspective, the majority of papers deal with KM-Tools.

De Vries and Brijder (2000) examine the key role of ICT to improve knowledge management in supply chain, since it contributes to the process of knowledge creation by providing four functions: adviser, assistant, librarian, and teacher. Malhotra et al. (2005) explore the nature and the composition of a knowledge management system analysing how collaboration between supply partners can lead to new knowledge creation in supply network, even when it may not be an explicit goal. The diversity of supply network partnership configurations are based on differences in capability platforms, reflecting varying processes and KM-Tools. This research study was conducted in the context of the *RosettaNet B2B* initiative, a consortium of major IT, electronic components, semiconductor manufacturing, telecommunications, and logistics enterprises working to create and implement open electronic business process standards for supply network collaboration. Malhotra et al. (2007) describe how the use of *SEBIs* (*standard electronic business interfaces*) improve the flexibility of supply network partnership and positively influence the adaptive knowledge creation process by enabling collaborative information exchange between supply firms. Zahay and Handfield (2004) show how KM-Tools contribute to knowledge storage and consist of four information-processing capabilities: generation, memory, dissemination, and interpretation. Using the web-based system *Ariba*, knowledge is stored in an integrated format inventory. Memory is defined as a sub-system to store knowledge for future use, developed with partners and based on "shoppable Internet catalog (SIC) capabilities". Rosu et al. (2009) suggest a knowledge based applications architecture based on the use of *enterprise resource planning (ERP), customer relationship management (CRM), document management system (DMS), data mining and data warehouse (DW)*. Lopez-Nicolas and Soto-Acosta (2010) identified *intranet* and *webpages* as KMS to support organisational learning. Al-Mutawah et al. (2009) propose a *multi-agent system (MAS)* for tacit knowledge transfer and perform some experiments to simulate the proposed approach. Similarly, Wu (2001) examines some specific multi-agent systems (*LivingFactory, DragonChain, StrategyFinder, eBAC*) used by supply firms to improve the knowledge transfer process. A multi-agent system is composed of a set of software programs that have the ability to integrate learned behaviour and work together towards some common goal. Even Goel et al. (2005) analyse how the use of specific multi-agent systems (*Farm Smart 2000, Heifer Management System, Casa*) and online auctions applications (Agriculture.com, Comdaq.net, Agex.com, Team.com, eBay.com) integrates the decision-making processes of different supply firms facilitating knowledge transfer and transparent economic transactions. Shih et al. (2012) describe the probable effects of a knowledge transfer system (*WEEKS*) in a complex environment. Findings of the case study indicate that the knowledge transfer process, when combined with adequate KM-Tools, could bridge the gaps between different partners with conflicting objectives. Huang and Lin (2010) analyse how current technologies, such as *electronic data interchange (EDI)* or the current *Web*, are useful for sharing data or information, rather than knowledge. A category of technologies to improve knowledge transfer process is the semantic web. Similarly, Douligeris and Tilipakis (2006) present a framework of *semantic ontologies* for knowledge transfer in supply network. The benefits provided by ontologies mainly deal with new ways of reducing the incompatibility problems of systems between firms and with new ways of transferring knowledge and sponsor products using a more semantic approach. Finally, Lin et al. (2012) analyse the use of a global *decision support system* used by suppliers to improve collaborative manufacturing processes. Levy et al. (2003) also suggest the importance of formal practices and methods (*manual formal, meetings with clients, formal process, formal meeting, send software meetings, report delivery*). Kovacs and Spens (2010) examine different types of *communities of practice (CoPs)* as a transfer practice for organised knowledge in the context of supply network. Wang et al. (2008) develop a conceptual-based model for knowledge transfer in supply networks based on the practice of *case-based reasoning (CBR)*. Chong et al. (2011) suggests a variety of knowledge transfer practices such as: *send employees to exhibitions/congress, send employees to universities/research institutes for further studies; use information from customers, suppliers or other organisations; establish strategy to obtain information from customers, suppliers, or others; hire know-how from advisors or consultants; use information from competitors to improve business performance; learning through customer-supplier partnership; learning through joint venture; learning through joint development agreement; learning through R&D contract; purchase licenses.*

In summary, the literature on KMS analyses specific web-based tools (Ariba, WEEKS, RosettaNet B2B, SEBIs, EDI), multi-agent systems (LivingFactory, DragonChain,, StrategyFinder, eBAC, Farm Smart 2000, Heifer Management System, Casa), online auctions applications (Agriculture.com, Comdaq.net, Agex. com, Team.com, eBay.com), decision support systems, semantic ontologies, knowledge transfer practices (CoPs, CBR, etc.). Nevertheless, they do not consider formal practices (casual mapping, knowledge mapping, balanced scorecard, formal manual, chief knowledge officer) and people-centred practices (brainstorming, focus groups, formal meetings, seminars, communities of sharing, informal networks, project teams, storytelling, job rotation, training). Then, literature does not propose a comprehensive taxonomy of knowledge management systems supporting supply firms in the different phases of knowledge management process and neither offers a more empirical evidence on their intensity of use. The above literature gap allows us to formulate the following research question:

RQ: How do KMS support supply firms in KM process?
In order to provide an answer to this research question a questionnaire survey has been carried out in a sample of supply firms. The following paragraph provides an overview of the methodology.

3.3 Methodology

This phase of field analysis provides as output the levels of the degree of adoption and the intensity of use of each KM-Tool and KM-Practice identified in the taxonomy provided by Cerchione and Esposito. These levels have been obtained through a questionnaire survey. The questionnaire used in this step has been developed into the following five phases:

1. *Definition of draft questionnaire* in which it has been prepared both a draft version of the questionnaire and the basic survey objectives;
2. *Meetings with experts* in which the two IT experts and one researcher have been again involved to test the suitability of the basic survey objectives and comprehensibility of the draft questionnaire;
3. *Definition of final questionnaire* in which survey objectives have been re-focused and the questionnaire has been revised on the basis of the feedback received during the meetings;
4. *Test of the questionnaire* in which the final version of the questionnaire has been tested through three pilot interviews carried out in three firms of the sample;
5. *Face-to-face distribution of questionnaire* in which the total number of respondents was 25 firms. As it is preferred for a qualitative problem as knowledge management in supply firms the questionnaire has been submitted by face-to-face mode to two managers with different skills and roles. This allowed us to obtain different strategic and operational perspectives.

In order to have a more comprehensive picture of the sample investigated, information from complementary sources (e.g. company websites, company reports, and industry magazines) have been collected and analysed.

The questionnaire survey has been carried out in a sample of 25 suppliers belonging in the south of Italy. This sample mainly comprises suppliers as shown in Table 3.1. In this table, the latest EU definition of SMEs proposed by the EU Commission has been used (European Commission 2005).

Table 3.1 Suppliers breakdown by employees' bands

Employees band	Number of firms	%
Micro 0–9	6	24
Small 10–49	16	64
Medium 50–249	3	12
Total	**25 firms**	

Tab. 3.2 Firm industries

Overall economic industry	Specific industry	Number of firms	%
Manufacturing	Aerospace	6	24
	Engineering	5	20
Service	Aerospace (R&D)	3	12
	ICT	9	36
	Management training and consulting	2	8
Total		**25 firms**	

Table 3.2 shows that the majority of suppliers investigated operate in high-tech industries even characterised by a high level of complexity, such as aerospace, engineering, ICT, management training, and consulting.

3.4 Results

This section describes the preliminary findings emerging from the field analysis. It is divided into two sub-sections. The first analysis shows the main results in terms of the degree of diffusion of KMS adopted by surveyed firms. The second describes the main findings regarding the intensity of use of KMS.

3.4.1 Degree of diffusion of KMS

The degree of diffusion of KM-Tools (KM-Practices) was defined as the number of firms adopting the specific KM-Tools (KM-Practices) divided by the total number of firms of the sample (25). It ranges from zero, if no SME uses the specific KM-Tools (KM-Practices), and 100, if all supply firms use the specific KM-Tool (KM-Practice).

As for KM-Tools, Fig. 3.1 highlights the degree of diffusion of KM-Tools (DT) which ranges from 12 (collaborative filtering, expert systems, crowdsourcing systems, mash up, syndication systems) to 96 (e-mail), with mean equal to 25.45, coefficient of variation 115.66 %. The high value of the coefficient of variation highlights that there is a focus only on few KM-Tools that are used by most supply firms investigated, while most KM-Tools are used only by a few supply firms or not used at all.

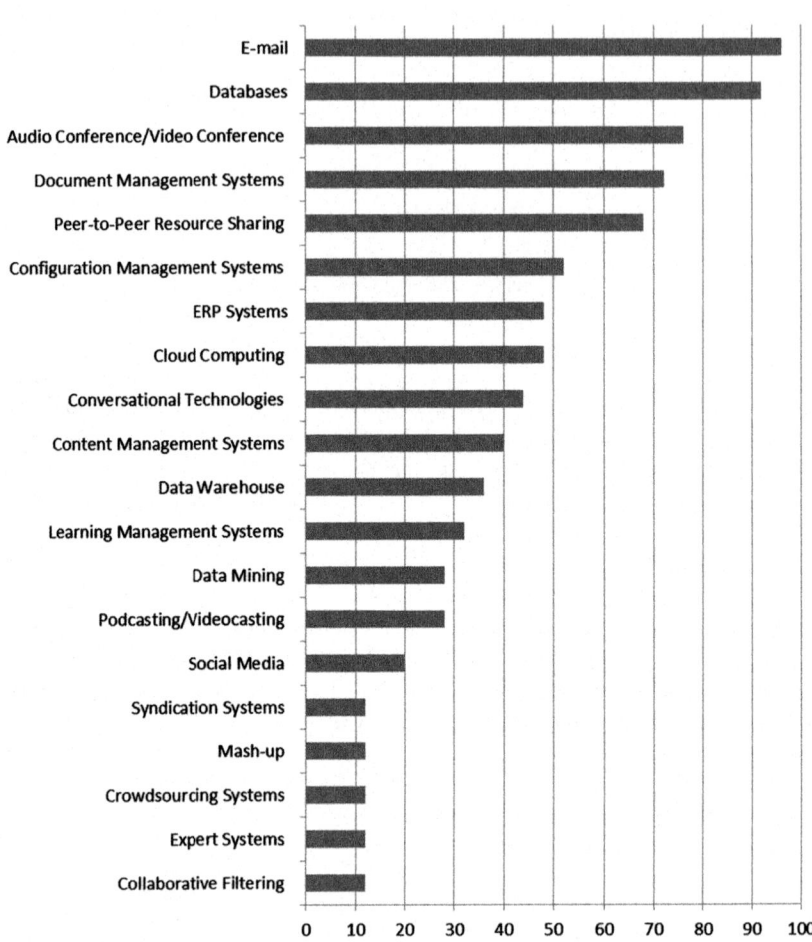

Fig. 3.1 KM-Tools—Levels for the degree of diffusion

These findings show that the suppliers investigated have perceived the strategic value of KM and consequently adopt a variety of KMS. Nevertheless, it emerged that firms are generally inclined to use not updated KM-Tools (e-mail, database, video conference/audio conference, document management systems, configuration management systems) instead of the newer ones (podcasting/videocasting, data mining, social media, mash-up, syndication systems, collaborative filtering, crowdsourcing systems), which are also cheaper and user friendly.

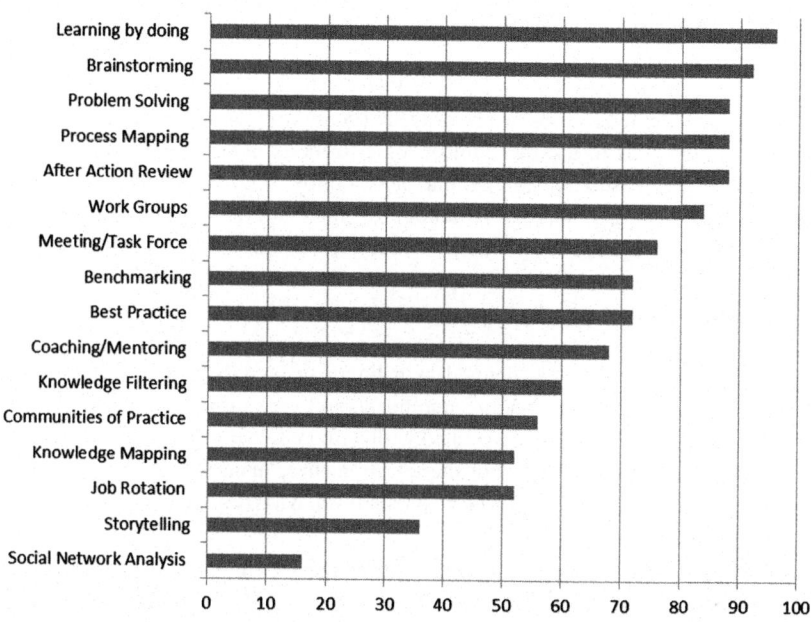

Fig. 3.2 KM-Practices—Levels for the degree of diffusion

With regard to KM-Practices, Fig. 3.2 highlights that the degree of diffusion is higher than that of M-Tools. Specifically, it ranges from 16 (social network analysis) to 96 (learning by doing), with mean equal to 33.21 and coefficient of variation 114.20%[1].

The high value of the coefficient of variation indicates that the supply firms investigated have a homogeneous behaviour with respect to the use of KM-Practices. In fact, most of KM-Practices are not used whereas the ones that are used are used by most supply firms investigated.

---

[1]Mean and coefficient of variation were calculated considering that the total number of KM-Practices is 33, even though 17 out of 33 KM-Practices identified are not used by any of the surveyed supply firms (ideas competition, knowledge elicitation interview, rating, casual mapping, knowledge modelling, balanced scorecard, contextual inquiry, knowledge office, lesson learned, case based reasoning, communities of sharing, focus groups, project teams training, facilitated discussion, informal networks, knowledge cafes, seminars).

3.4.2 Intensity of use of KMS

To evaluate the intensity of use of each KM-Tool and KM-Practice a fuzzy set theory based approach was used (Zadeh 1965; Watanabe 1979). The fuzzy set theory allows us to use the rigor of logic to model natural language and the common sense reasoning (Michellone and Zollo 2000; Zimmermann 2001). Therefore, it is an adequate methodology to aggregate approximate judgments expressed by managers during the face-to-face survey implementation. In particular, the intensity of use of KM-Tools (KM-Practices) was calculated as described in the following six steps:

1. The intensity of use was defined as a linguistic variable on five qualitative levels: very poor, poor, medium, significant, and very significant.
2. Each qualitative level was associated with a fuzzy set (Fig. 3.3).
3. During the phase of filling out the questionnaire, managers provide a qualitative judgment about the level of intensity of KM-Tools (KM-Practices) by their firms.
4. Each qualitative judgment was codified into the correspondent fuzzy set.
5. For each KM-Tool (KM-Practice), the fuzzy mean was calculated.
6. The fuzzy mean was de-fuzzified using the mean-of-maxima (MeOM) technique. The results representing the intensity of use of an individual KM-Tool (KM-Practice) are presented into two paragraphs, depicted in Fig. 3.3.

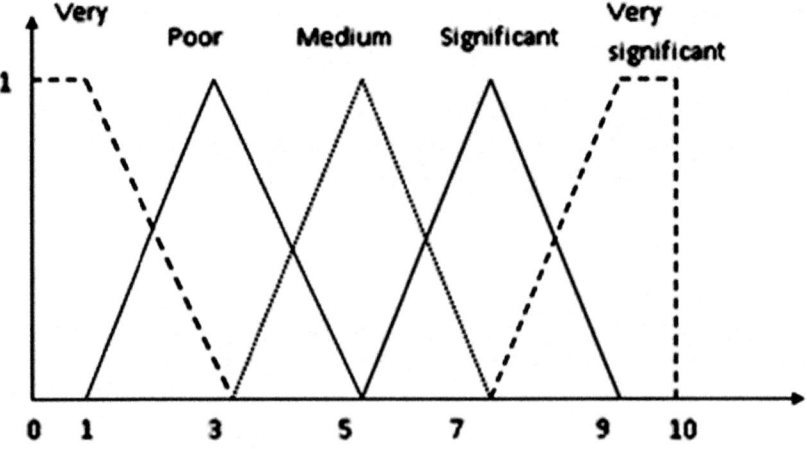

Fig. 3.3 Fuzzy sets representing qualitative judgements for the intensity of use of KMS

As for KM-Tools, Fig. 3.4 highlights that the intensity of use of KM-Tools (IT) ranges from 13 (collaborative filtering) to 95 (e-mail), with a mean equal to 49.70 and coefficient of variation 42.72. The low value of the coefficient of variation indicates that the intensity of use of KM-Tools is quite homogeneous. Most KM-Tools have an intensity of use around the mean, whereas few of them have high or low intensity of use.

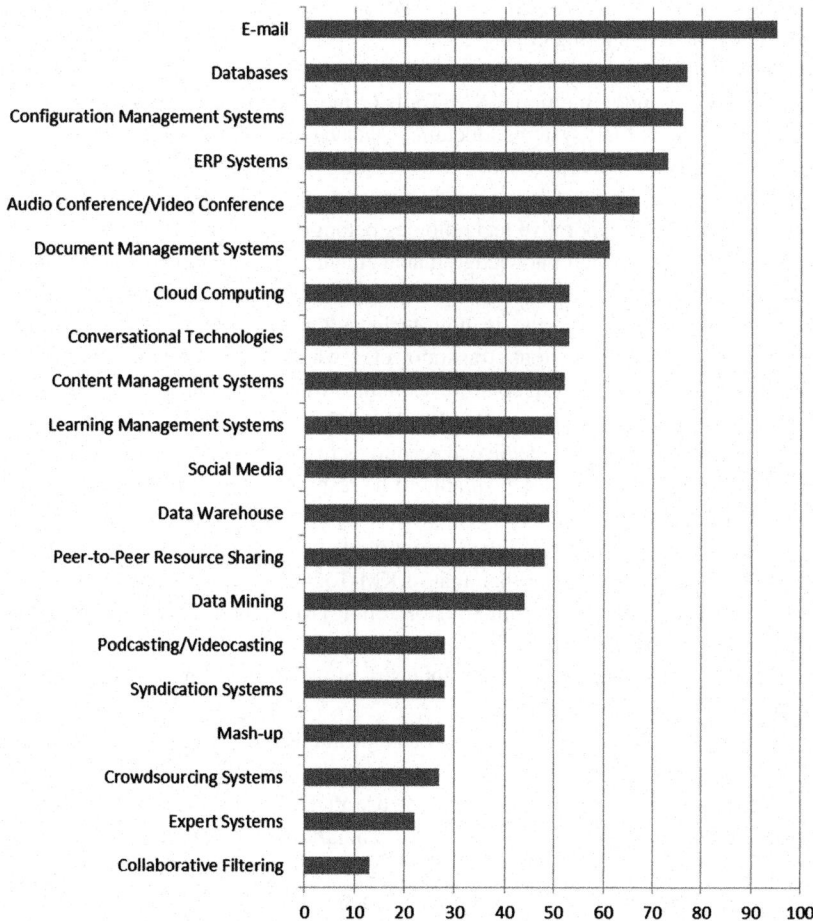

Fig. 3.4 KM-Tools—Levels for the intensity of use

A first group of KM-Tools with a high intensity of use includes e-mail (95), databases (77), configuration management systems (76), ERP systems (73), audio conference/video conference (67), and document management systems (61). A second group of KM-Tools with a medium intensity of use includes cloud computing (53), conversational technologies (53), content management systems (52), learning management systems (50), social media (50), data warehouse (49), peer-to-peer resource sharing (48), data mining (44). Finally, a third group of KM-Tools with a low intensity of use includes podcasting/videocasting (28), syndication systems (28), mash-up (28), crowdsourcing systems (27), expert systems (22), collaborative filtering (13).

These findings show that the supply firms investigated are generally inclined to use intensely more traditional KM-Tools (e-mail, databases, audio conference/video conference, ERP systems, document management systems) instead of new and more updated tools (podcasting/videocasting, mash-up, syndication systems, collaborative filtering) that, as stressed above, are generally cheaper and easier to use. Specifically, collaborative technologies belonging to Web 2.0 are not intensively used to improve the knowledge management process in terms of efficiency and effectiveness. This aspect is even more significant when considering that the supply firms analysed operate in high-tech and/or complex industries, such as aerospace, telecommunications, transport, etc., where large companies adopt the most updated KMS. This gap could be explained by the rapid technological changes in the ICTs industry represented by Web 2.0. Firms and in particular supply firms typically do not have dedicated resources to monitor and follow the evolution of Web 2.0. They are not even able to be responsive to the technology dynamics. This forces them to remain in a backward position.

Concerning KM-Practices, Fig. 3.5 highlights that the intensity of use of KM-Practices is on average higher than that of KM-Tools. Specifically, the intensity of use of KM-Practices ranges from 36 (storytelling) to 73 (problem solving), with a mean equal to 52.75 and coefficient of variation 19.72. The low value of the coefficient of variation indicates that the intensity of use of different KM-Practices is homogeneous.

A first group of KM-Practices with a high intensity of use includes problem solving (73), learning by doing (71), meeting/task force (62). A second group of KM-Practices with a medium intensity of use includes work groups (58), brainstorming (58), process mapping (56), benchmarking (56), after action review (52), knowledge mapping (50), best practice (50), coaching/mentoring (49), knowledge filtering (48), communities of practice (45). Finally, a third group of KM-Practices with a low intensity of use includes social network analysis (40), job rotation (40), and storytelling (36).

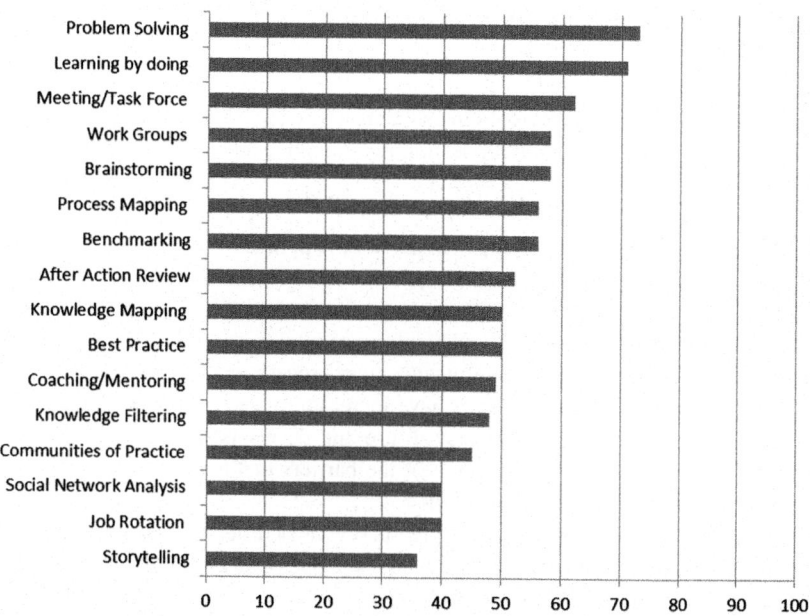

Fig. 3.5 KM-Practices—Levels for the intensity of use

As for the *RQ* presented at the end of Sect. 3.2, this chapter highlights in summary that the majority of firms investigated adopts a variety of more traditional KM-Tools instead of new and more updated tools that are generally cheaper and easier to use. During the interviews, interviewees have underlined that this gap is a consequence of two factors. On the one hand, supply firms typically do not have dedicated resources to monitor the evolution of the ICT market and are not even able to follow the technological dynamic. This forces them to remain in a backward position. Therefore, this gap highlights the difficulties in following rapid technological changes and the lack of support from the ICT providers.

These results show that the investigated supply firms adopt intensely traditional practices not focused exclusively on knowledge management issues (problem solving, learning by doing, meeting/task force, work groups) instead of new practices (communities of practice, knowledge filtering, knowledge mapping) that are more centred on the KM process. Nevertheless, it emerges from the interviews that this conclusion is strictly connected with the nature of knowledge which in these firms is prevalently human embedded and does not promote a large diffusion of formal KM-Practices.

3.5   Conclusions and implications

The main aim of this chapter is to contribute to an increase of the body of knowledge in the field of KM in supply management. The analysis of the literature has allowed us to identify a research question regarding the adoption of KMS in supply firms:

RQ: How do KMS support supply firms in KM process?
To provide an answer to this research question, this chapter presents a taxonomy of KM-Tools and KM-Practices. These two taxonomies offer to supply firms the opportunity to identify a set of tools and practices that can be used to improve the different phases of knowledge management process (creation, storage, transfer). In fact, nowadays supply firms have increasingly access to new knowledge management systems, which are easy to use and do not need significant investments. This has allowed the reduction of the barriers that hindered the spread of knowledge management in supply firms.

With these premises, a questionnaire survey carried out on a sample of suppliers operating in high-tech and/or complex industries. The findings highlight that the majority of firms investigated adopts a variety of more traditional KM-Tools instead of new and more updated tools that are generally cheaper and easier to use. During the interviews, interviewees have underlined that this gap is a consequence of two factors. On the one hand, supply firms typically do not have dedicated resources to monitor the evolution of the ICT market and are not even able to follow the technological dynamic. This forces them to remain in a backward position. Therefore, this gap highlights the difficulties in following rapid technological changes and the lack of support from the ICT providers.

In addition, the results show that the investigated supply firms adopt intensely traditional practices not focused exclusively on knowledge management issues (problem solving, learning by doing, meeting/task force, work groups) instead of new practices (communities of practice, knowledge filtering, knowledge mapping) that are more centred on KM process. Nevertheless, it emerges from the interviews that this conclusion is strictly connected with the nature of knowledge which in these firms is prevalently human embedded and does not promote a large diffusion of formal KM-Practices. Practices can be learned from customers or suppliers in spite of tools. This aspect demonstrates how suppliers may improve their business through subcontracting strategies.

The survey results provide guidance for future research. The first research implication derives from the fact that supply firms generally use not updated KMS instead of the newer ones. This issue requires further and in-depth analysis concerning the degree of alignment between KMS used by supply firms and the nature of knowledge from both the ontological and epistemological perspectives. Secondly, due to the increasing importance of firm networks in the development of supply firms, it seems important to investigate the ways through which knowledge is spread across networks.

From the supply firms point of view, this paper has highlighted that supply firms could further increase the impact of KM by better exploiting the opportunity offered by the new ICTs (such as cloud computing, crowdsourcing system, collaborative filtering, wiki, etc.).

From a KMS providers point of view, this paper has stressed that supply firms typically do not have dedicated resources to monitor the process of innovation in the field of the KMS. Nevertheless, they could represent a significant market. To catch this opportunity, it is necessary to not only create a new market segment dedicated to supply firms but also direct channels of communication (even virtual) between supply firms and KMS providers.

References

Abid, M., Ali, B.: Antecedents and effectiveness of CKM: an empirical study. Middle-East. J. Sci. Res. 19(7), 880–892 (2014)

Alavi, M., Leidner, D.L.: Review: knowledge management and knowledge management systems: conceptual foundations and research issues. MIS Quart. 25(1), 107–136 (2001)

Al-Mutawah, K., Lee, V., Cheung, Y.: A new multi-agent system framework for tacit knowledge management in manufacturing supply chains. J. Intell. Manuf. 20(5), 593–610 (2009)

Blome, C., Schoenherr, T., Eckstein, D.: The impact of knowledge transfer and complexity on supply chain flexibility: a knowledge-based view. Int. J. Prod. Econ. 147, Part B, 307–316 (2014)

Chang, C.W., Chiang, D.M., Pai, F.Y.: Cooperative strategy in supply chain networks. Ind. Market. Manag. 41,1114–1124 (2012)

Cheng, J.H., Fu, Y.C.: Inter-organizational relationships and knowledge sharing through the relationship and institutional orientations in supply chains. Int. J. Inform. Manag. 33, 473–484 (2013)

Chong, C.W., Chong, S.C., Gan, G.C.: Inter-organizational knowledge transfer needs among small and medium enterprises. Libr. Rev. 60(1), 37–52 (2011)

Corso, M., Martini, A., Pellegrini, L., Paolucci, E.: Technological and organizational tools for knowledge management: in search of configurations. Small Bus. Econ. 21(4), 397–408 (2003)

De Vries, E.J., Brijder, H.G.: Knowledge management in hybrid supply channels: a case study. Int. J. Technol. Manag. 20(5–8), 569–587 (2000)

Desouza, K.C., Chattarai, A., Kraft, G.: Supply chain perspective to knowledge management: research propositions. J. Knowl. Manag. 7, 129–138 (2003)

Douligeris, C., Tilipakis, N.: A knowledge management paradigm in the supply chain. EuroMed J. Bus. **1**, 66–83 (2006)

European Commission, The New SME Definition. User Guide and Model Declaration. European Commission, DG Enterprise & Industry, Bruxelles (2005)

Fugate, B.S., Autry, C.W., Davis-Sramek, B., Germain, R.N.: Does knowledge management facilitate logistics-based differentiation? The effect of global manufacturing reach. Int. J. Prod. Econ. **139**(2), 496–509 (2012)

Garrigos-Simon, F.J., Lapiedra Alcami, R., Barbera Ribera, T.: Social networks and Web 3.0: their impact on the management and marketing of organizations. Manag. Decis. **50**(10), 1880–1890 (2012)

Goel, A., Zobel, C.W., Jones, E.C.: A multi-agent system for supporting the electronic contracting of food grains. Comput. Electron. Agric. **48**, 123–137 (2005)

Halley, A., Nollet, J., Beaulieu, M., Roy, J., Bigras, Y.: The impact of the supply chain on core competencies and knowledge management: direction for future research. Int. J. Technol. Manage. **49**, 297–313 (2010)

He, Q., Ghobadian, A., Gallear, D.: Knowledge acquisition in supply chain partnerships: the role of power. Int. J. Prod. Econ. **141**(2), 605–618 (2013)

Huang, C.C., Lin, S.H.: Sharing knowledge in a supply chain using the semantic web. Expert Syst. Appl. **37**, 3145–3161 (2010)

Hult, G.T.M., Ketchen, D.J., Cavusgil, T., Calantone, R.J.: Knowledge as a strategic resource in supply chains. J. Oper. Manag. **24**, 458–475 (2006)

Hult, G.T.M., Ketchen, D.J., Slater, S.F.: Information processing, knowledge development, and strategic supply chain performance. Acad. Manag. J. **47**, 241–253 (2004)

Kanat, S., Atilgan, T.: Effects of knowledge management on supply chain management in the clothing sector: Turkish case. Fibres Text. East. Eur. **103**, 9–13 (2014)

Kim, K.K., Umanath, N.S., Kim, J.Y., Ahrens, F., Kim, B.: Knowledge complementarity and knowledge exchange in supply channel relationships. Int. J. Inform. Manag. **32**(1), 35–49 (2012)

Kovacs, G., Spens, M.K.: Knowledge sharing in relief supply chain. Int. J. Network. Virtual Organ. **7**, 222–239 (2010)

Levy, M., Loebbecke, C., Powell, P.: SMEs, co-opetition and knowledge sharing: the role of information systems. European J Inform. Syst. **12**(1), 3–17 (2003)

Lin, I., Seidel, R., Shekar, A., Shahbazpour, M., Howell, D.: Knowledge sharing differences between engineering functional teams: An empirical investigation. J. Inform. Know. Manag. **11**(3), 1–14 (2012)

Liu, Y., Huang, Y., Luo, Y., Zhao, Y.: How does justice matter in achieving buyer-supplier relationship performance? J. Oper. Manag. **30**, 355–367 (2012)

Loke, S.P., Downe, A.G., Sambasivan, M., Khalid, K.: A structural approach to integrating total quality management and knowledge management with supply chain learning. J. Bus. Econ. Manag. **13**(4), 776–800 (2012)

Lopez-Nicolas, C., Soto-Acosta, P.: Analyzing ICT adoption and use effects on knowledge creation: An empirical investigation in SMEs. Int. J. Inform. Manag. **30**(6), 521–528 (2010)

Lu, Q., Meng, F., Goh, M.: Choice of supply chain governance: Self-managing or outsourcing? Int. J. Prod. Econ. **154**, 32–38 (2014)

Mak, K.T., Ramaprasad, A.: Knowledge supply network. J. Oper. Res. Soc. **54**(2), 175–183 (2003)

Malhotra, A., Gosain, S., El Sawy, O.A.: Absorptive capacity configurations in supply chains: Gearing for partner-enabled market knowledge creation. MIS Quart. **29**(1), 145–187 (2005)

Malhotra, A.A., Gosain, S.B., Sawy, O.A.E.C.: Leveraging standard electronic business interfaces to enable adaptive supply chain partnerships. Inform. Syst. Res. **18**(3), 260–279 (2007)

Michellone, G., Zollo, G.: Competences management in knowledge-based firms. Inter. J. Technol. Manag. **20**(1), 134–135 (2000)

Rosu, S.M., Dragoi, G., Guran, M.: A knowledge management scenario to support knowledge applications development in small and medium enterprises. Adv. Electr. Comput. Eng. **9**(1), 8–15 (2009)

Sambasivan, M., Loke, S.P., Abidin-Mohamed, Z.: Impact of knowledge management in supply chain management: A study in Malaysian manufacturing companies. Know. Proc. Manag. **16**(3), 111–123 (2009)

Shih, S.C., Hsu, S.H.Y., Zhu, Z., Balasubramanian, S.K.: Knowledge sharing-A key role in the downstream supply chain. Inform. Manag. **49**(2), 70–80 (2012)

Wang, C., Fergusson, C., Perry, D., Antony, J.: A conceptual case-based model for knowledge sharing among supply chain members. Bus. Proc. Manag. J. **14**, 147–165 (2008)

Watanabe, N.: Statistical methods for estimating membership functions. Japanese J. Fuzzy Theory Syst **5**(4), (1979)

Wu, D.J.: Software agents for knowledge management: coordination in multi-agent supply chains and auctions. Expert Syst. Appl. **20**, 51–64 (2001)

Zadeh, L.A.: Fuzzy Sets. Inform. Control **8**(3), 338–353 (1965)

Zahay, D.L., Handfield, R.B.: The role of learning and technical capabilities in predicting adoption of B2B technologies. Ind. Market. Manag. **33**(7), 627–641 (2004)

Zhang, Q.J., Zhou, K.Z.: Governing interfirm knowledge transfer in the Chinese market: The interplay of formal and informal mechanisms. Ind. Market. Manag. **42**(5), 783–791 (2013)

Zimmermann, H.: Fuzzy Set Theory and its Applications. Kluwer Academic Publishers, Boston (2001)

About the Authors

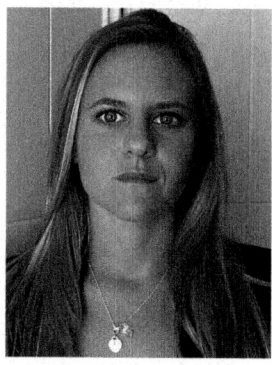

Piera Centobelli received with honors the M.Sc. Degree in Engineering Management at the University of Naples Federico II. From 2013 until 2016 Centobelli joined the Fraunhofer J-LEAPT. In 2016 she awarded a Ph.D. in Technology and Production Systems at the Department of Chemicals, Materials and Industrial Production Engineering by the same university. Currently she is Postdoctoral Research Fellow at the University of Naples Federico II. Her research interests focus on operations management, logistics and supply chain management, big data and analytics in logistics and sustainable supply chain management, knowledge and technology management, and development of decision support systems.

Roberto Cerchione is Aggregate Professor of Business Management at the Faculty of Engineering and Assistant Professor of Engineering Management at the Department of Engineering of the University of Naples Parthenope. He received with honors the M.Sc. Degree in Engineering Management and awarded a Ph.D. in Science and Technology Management by the University of Naples Federico II. His research projects are focused on knowledge and technology management, big data and analytics, supply chain management and environmental sustainability management in high-tech manufacturing and service industries.

Emilio Esposito is Professor of Engineering Management at the Department of Industrial Engineering of the University of Naples Federico II. He was awarded a Ph.D. in Economics of Technological Change by the University of Naples Federico II. His current scientific interests include knowledge management in SMEs, technology management, supply chain management, industrial organization in high tech industries.

The effect of digitalization is measurable – How transparency on the CCC creates momentum for working capital management?

4

Lotta Lind, Timo Kärri, Sari Monto, Miia Pirttilä und Florian Schupp

Abstract

The management of working capital and financial supply chains has recently gained increasing interest in companies as well as in academia. The advantages of digitalization are diverse in this area, and there is remarkable potential to release tied-up working capital through collaborative actions in the value chains. In this chapter, we review selected literature on measuring and managing inter-organizational working capital, and discuss the effect of digitalization on working capital management. Lastly, we propose a concept for measuring, monitoring, and managing working capital at the value chain level.

L. Lind (✉) · T. Kärri · S. Monto · M. Pirttilä
Lappeenranta, Finnland
E-Mail: lotta.m.lind@gmail.com

T. Kärri
E-Mail: timo.karri@lut.fi

S. Monto
E-Mail: sari.h.monto@gmail.com

M. Pirttilä
E-Mail: miia.pirttila@lut.fi

F. Schupp
Bühl, Deutschland
E-Mail: schupp-florian@t-online.de

© Springer Fachmedien Wiesbaden GmbH 2018
F. Schupp und H. Wöhner (Hrsg.), *Digitalisierung im Einkauf*,
https://doi.org/10.1007/978-3-658-16909-1_4

The main element of the concept is a web-based platform which enables the linking and managing of the physical and financial flows in the value chain via modern technologies and real-time information flow.

4.1   Background

Working capital management (WCM) has become a „hot topic" recently. Both academics and practitioners have realized the importance of effective working capital management and searched potential objects to release capital through more effective management of inventories and trade credit. In academic journals, the number of published articles on working capital management has increased widely in recent years: during 1990–2010, 23 journal articles concerning working capital management in the studied databases were found (Viskari et al. 2011a), whereas the number in the 3-year-period of 2011–2013 only was as much as 39 articles (Pirttilä 2014). Also companies, such as BMW (2010) and Valeo (2012) in the automotive industry, have highlighted the importance of working capital management in their annual reports. At Valeo, close supplier integration has even enabled a financial model based on a negative working capital (Valeo 2012). The reason behind the strongly increased interest towards working capital management dates back to the year 2008 and the financial crisis which led to the deteriorated general financial situation as well as tightened possibilities to get external funding. The situation made companies focus on working capital management, but also created demand for capital from the supply chain (Hofmann et al. 2011). Lately, many studies have emphasized the inter-organizational approach to working capital management which looks at working capital from the perspective of a value chain instead of an individual company (e.g. Hofmann and Kotzab 2010; Grosse-Ruyken et al. 2011; Lind et al. 2012; Viskari et al. 2012a; Wuttke et al. 2013). In addition to releasing capital for more strategic objectives, the motive for paying attention to working capital management in companies may be the negative correlation between working capital management and profitability, which has been indicated by many studies, i.e. companies can improve their profitability by reducing the tied-up working capital (e.g. Jose et al. 1996; Shin and Soenen 1998; Deloof 2003; Lazaridis and Tryfonidis 2006; Talha et al. 2010; Enqvist et al. 2014).

Digitalization has been changing the world since the 1980s, when computers in households started to become more common. With the fast development of technologies, digitalization has provided new opportunities and ways for doing business. Currently, we are shifting from the third to the fourth industrial revolution: from automating production via use of electronic and IT systems to the use of cyber-physical systems and smart factories (Germany Trade & Invest 2014). The advantages of digitalization are diverse in the management of financial supply chains. Accurate information about inventory levels could be used for more efficient inventory management and could even be shared with the value chain partners to improve the physical flows of material and performance of the whole chain. Today's technological solutions would enable radical updates to current payment term practices adopted after an era of cash payments. Hence, there is an enormous potential to reduce tied-up working capital and improve the efficiency of the financial supply chains through digitalization.

In this chapter, the effect of digitalization on working capital management is discussed. Figure 4.1 illustrates the framework for the chapter. First, we introduce measuring working capital via cycle times. After that, some methods and tools for analyzing working capital are presented, and selected previous research is reviewed. This is followed by a section where the effect of digitalization on working capital management and its components is discussed, and after that a concept for a working capital management platform for a value chain is suggested. Finally, the conclusions and suggestions for future research are presented.

4.2 Measuring working capital by Cash Conversion Cycle (CCC)

Time has been recognized as a source of competitive advantage (Gehani 1995; Ng et al. 1997) and it has been widely used as a measure in supply chain management. In this chapter, the management of working capital and its components is measured by cycle times. Cash conversion cycle (CCC), the cycle time of working capital, developed by Richards and Laughlin (1980), combines the working capital components and describes the number of days a firm has funds tied up in operational working capital. The shorter the CCC, the less external funding is needed to finance the working capital. The cash conversion cycle is defined as follows:

CCC = DIO + DSO – DPO,
with
DIO = Days inventories outstanding = (Inventories × 365)/sales
DSO = Days sales outstanding = (Accounts receivable × 365)/sales
DPO = Days accounts payable outstanding = (Accounts payable × 365)/sales

Figure 4.2 illustrates the positive cash conversion cycle starting from a purchase from the supplier and ending with the payment received from the customer. A negative

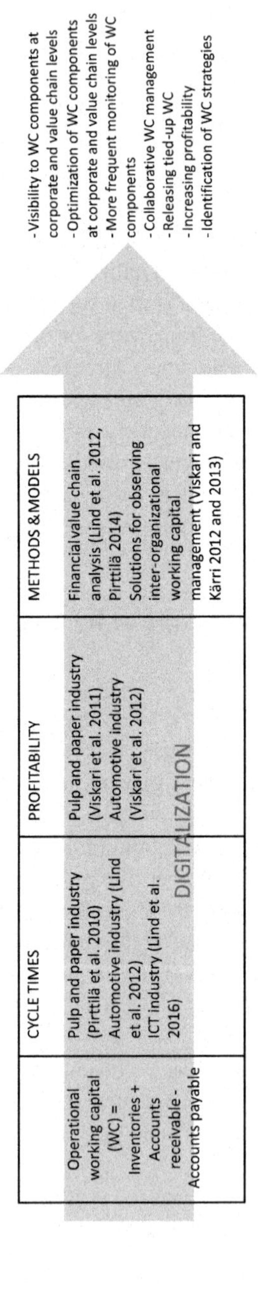

Fig. 4.1 Framework of the chapter

The figure shows a large upward-pointing arrow labeled DIGITALIZATION. Below the arrow is a table with the following columns:

	CYCLE TIMES	PROFITABILITY	METHODS & MODELS
Operational working capital (WC) = Inventories + Accounts receivable - Accounts payable	Pulp and paper industry (Pirttilä et al. 2010) Automotive industry (Lind et al. 2012) ICT industry (Lind et al. 2016)	Pulp and paper industry (Viskari et al. 2011) Automotive industry (Viskari et al. 2012)	Financial value chain analysis (Lind et al. 2012, Pirttilä 2014) Solutions for observing inter-organizational working capital management (Viskari and Kärri 2012 and 2013)

At the top of the arrow:
- Visibility to WC components at corporate and value chain levels
- Optimization of WC components at corporate and value chain levels
- More frequent monitoring of WC components
- Collaborative WC management
- Releasing tied-up WC
- Increasing profitability
- Identification of WC strategies

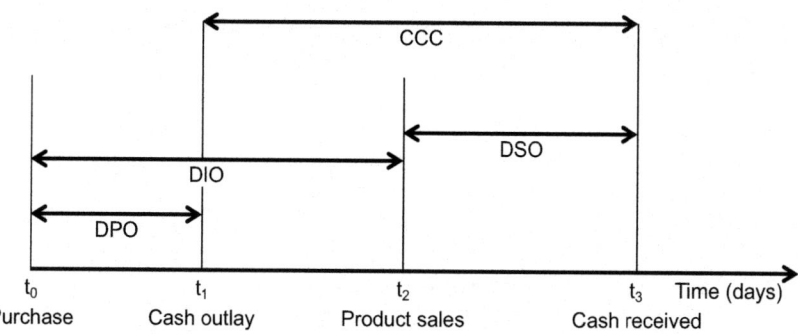

Fig. 4.2 Cash conversion cycle. *Source* adapted from Richards and Laughlin (1980)

cash conversion cycle can be achieved when the cycle time of accounts payable is so long that the company receives the payment from the customer before it pays for its raw material supplier. This way, the operations of the company are financed by the customer and the supplier. A negative CCC is exceptional and rarely achieved, but for example the companies Apple and Dell are applying working capital strategies which enable situations where no working capital is tied-up in the company itself (Lind et al. 2016).

The cash conversion cycle has also been developed further as well as used in different tools. The adjusted cash conversion cycle (ACCC) is an internal information tool for controlling the amount and cost of working capital in the internal value chain (Viskari et al. 2012b). The modified cash conversion cycle (mCCC) also takes into account the received advance payments, which are common for example in the project business and ICT industry, as a component of operational working capital (Talonpoika et al. 2014). Viskari and Kärri (2012, 2013) have developed tools for measuring and controlling working capital in an inter-organizational context. The efficiency of working capital management and the company's position in the value chain can be observed through a model which measures two cycle times: the CCC and ACCC, in addition to the financing costs of working capital (Viskari and Kärri 2012). This tool is used at corporation level in the inter-organizational value chain. The results in the value chain of four automotive companies suggested that financing costs decrease by shortening the cycle times of inventories as well as the cycle times of accounts payable and receivable, but reducing the inventories was more efficient. The financial cycle time model (FCTM) (Viskari and Kärri 2013) is a practical tool for observing tied-up working capital in the value chain of a product. Scenarios showed that the shortening of the cycle times of working capital can have a positive effect on performance.

4.3   How can the effects of digitalization be measured with CCC?

In this section, some methods and tools for analyzing working capital management are introduced. The results of the selection of previous studies are also shortly reviewed.

4.3.1   Financial value chain analysis

The majority of the previous studies on working capital management focus on individual companies. Only recently the research stream of financial supply chain management has emerged and brought the physical and financial flows into discussion from the perspective of the entire value chain. The efficient management of upstream flow of money is considered as important as the management of downstream flow of goods (Gupta and Dutta 2011). In the current business environment, where competition is based more and more on the efficiency of the value chains rather than individual companies, working capital management is also a topic that should be examined from a more holistic perspective. It is not even possible for one company to manage its working capital alone, as the management of financial flows is affected by the actions of its suppliers and customers as well. Holistic, supply-chain oriented view to working capital management has been emphasized by many authors (e.g. Hofmann and Kotzab 2010; Viskari et al. 2012a; Seifert et al. 2013; Lorentz et al. 2016; Protopappa-Sieke and Seifert 2016). Studies have also found evidence that the working capital optimization of one company by payment term adjustments at the expense of the value chain partners only leads to short-term success (Grosse-Ruyken et al. 2011; Huff and Rogers 2015).

Financial value chain analysis (Lind et al. 2012; Pirttilä 2014) is a systematic method for analyzing the financial aspects in the value chain. The method enables the analysis of the value chain, consisting of the companies operating in the different stages of the chain, by several ways: the tool reveals the positions of the stages in the value chain as well as the positions of the companies within the stages, but also comparisons between the performances of different value chains can be made. The financial value chain analysis method consists of seven steps which are: (1) Choose the industry to be studied, (2) Define the value chain, including the stages and companies in the chosen industry, (3) Define what the key figures to be studied are, (4) Collect data (preferably from official and public sources), (5) Calculate the values for the selected key figures, (6) Analyze the key figures, and (7) Draw conclusions. The financial value chain analysis has been used to study working capital management by cycle times in the automotive industry (Lind et al. 2012), pulp and paper industry (Pirttilä et al. 2010, 2014) and information and communication technology (ICT) industry (Lind et al. 2016). The findings indicated that the relationship between sales and working capital is nearly constant at the value chain level, as there were no remarkable changes in the average cycle times of working capital of the value chains during the observation periods. There were differences in the CCCs between the industries: the value chain of the ICT industry had the shortest average CCC, ca. 40 days, while the CCCs of the value chain of the automotive and pulp and paper industries were around 70 and 60 days, respectively. The differences between the industries were mainly due to the differences in the cycle times of inventories.

The study in the automotive industry (Lind et al. 2012) included 65 companies operating in six stages in the value chain from raw material suppliers to car dealers. Even if the CCC did not change during the observation period of 2006–2008, the cycle times of accounts receivable (DSO) and accounts payable (DPO) had remarkable changes, while the cycle time of inventories (DIO) remained on almost the same level. However, these changes in the DSO and DPO did not affect the CCC, as the changes in the financial flows, DSO and DPO, offset each other in the value chain context. This finding highlights the important role of inventory management in working capital management. The value chain of the pulp and paper industry (Pirttilä et al. 2010, 2014) consisted of approximately 40 companies located in eight different stages. The results from the observation period of 2004–2008 indicated that the CCC is shorter in the downstream part of the chain, whereas more capital-intensive companies in the upstream tied up more working capital. The ICT industry, in turn, had some exceptions in working capital management in comparison to the automotive and pulp and paper industries (Lind et al. 2016). In addition to a notably shorter average CCC of the sample of 61 companies, several companies were found to be operating with a negative CCC. The inventories were effectively managed in all nine stages during the observation period of 2006–2010, but companies with a negative CCC were having a long cycle time of accounts payable (DPO) as well.

4.3.2   Working capital management and profitability

The relation between working capital management and profitability has been studied widely, and it is the most studied topic in the literature of working capital management. Several studies have shown the negative correlation between working capital management and relative profitability (e.g. Jose et al. 1996; Shin and Soenen 1998; Deloof 2003; Lazaridis and Tryfonidis 2006; García-Teruel and Martínez-Solano 2007; Talha et al. 2010; Enqvist et al. 2014). However, many of these studies have used large, multi-industry datasets and considered profitability from the perspective of an individual company. Figure 4.3 illustrates working capital management in the value chain context and describes the relationship between working capital management and profitability.

Viskari et al. (2011b, 2012a) studied the connection between working capital components and relative profitability (ROC%) in the value chains of the pulp and paper and automotive industries. In the pulp and paper industry, the results supported the previous findings of the negative relation between working capital management and profitability, but in the automotive industry the results suggested that, depending on their position in the value chain, companies benefit from different working capital strategies and not only from shortening the CCC to a

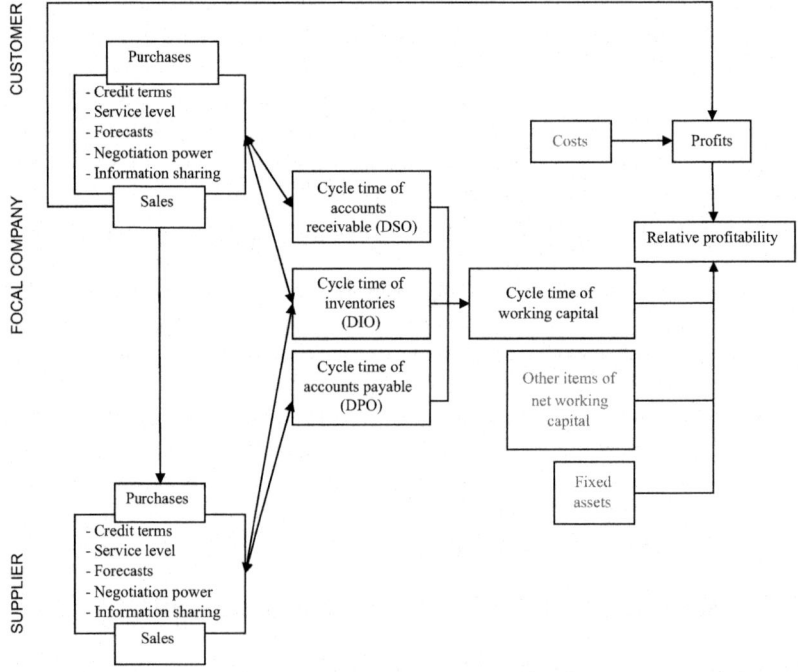

Fig. 4.3 Impact of working capital management on profitability in the value chain context. *Source* Viskari et al. (2012a)

minimum. Simulations suggested that the profitability of the value chain could be improved either by managing all working capital components simultaneously, or by radical reduction of payment terms.

4.3.3 Models for working capital management

Mullins and Komisar (2009) define the working capital model as a part of the company's business model. As found by Viskari et al. (2012a), all companies in the value chain do not benefit from similar practices related to working capital management, but have different models for managing working capital. In the value chain context, it would not even be possible for all companies to reduce working capital to zero due to different premises regarding their position in the value chain, bargaining power and financial conditions. In addition, the actions in the value chain affect other companies as well, and therefore the improvements in working capital management should not be done at the expense of other companies by passing the negative effects to suppliers or customers, as this would weaken the competitiveness of the chain.

Monto et al. (2013) studied working capital models in the value chain of the automotive industry by statistical cluster analysis. Four different working capital models were found: the Successful minimizing model had short cycle times of all working capital components, whereas the Inventory holding model differed from others by having a long cycle time of inventories (DIO), which was compensated with a relatively short cycle time of accounts receivable (DSO). The Aiming-at-minimum model had not been able to shorten the cycle times of working capital as much as the successful minimizers. The fourth model was the Credit granting model, which consisted of companies having a significantly longer DSO than DIO. The results also indicated that different parts of the value chain have typical working capital models. This was also found in the study of the ICT industry (Lind et al. 2016), but it was noted that even if there is a dominating working capital model for a value chain stage, not all companies within the stage operate with the same working capital model.

Identifying the working capital management models applied by companies in the value chain is a prerequisite for the optimization of working capital management at the value chain level. Each firm should ensure that their cycle times are in line with the structure of the value chain, and the positions in the value chain should be understood before making decisions about working capital management (Grosse-Ruyken et al. 2011; Wuttke et al. 2013). Analysis of working capital models applied and internal decisions of which strategy to follow are the starting point for a value chain's working capital management optimization.

4.4   How does digitalization affect the CCC and its components?

The advantages of digitalization offer many opportunities for making working capital management in the value chains more effective. Releasing working capital in the value chains would benefit both the companies and the value chains. The visibility of real-time information about the working capital components would enable the management of the CCC and its components in a more accurate manner. However, managing working capital in the supply chain is challenging due to the fact that there is no single person or board of managers responsible for the entire supply chain. Companies should trust each other and commit to the collaborative targets regarding working capital management agreed together with the supply chain partners. This requires new management practices and tools for collaboration, such as working capital pools (Protopappa-Sieke and Seifert 2016), and it should be considered which tools for different levels of collaboration should be used, as openness and trust in the value chains require long-term relationships (Ali-Marttila et al. 2016). Studies have shown that there is potential to reduce the CCC to remarkably shorter values. Of course, the old patterns regarding working capital management, e.g. payment terms, cannot be changed in the whole industry at once, but some companies and value chains should take on the role of trailblazers.

In the management of inventories, remarkable potential lies in making the inventories visible between the value chain partners. The calculation of accurate order points, automated order processes and replenishments, and the possibility to view the inventory levels of customers and suppliers could reduce inventory levels throughout the entire chain. Information sharing and automated processes would be key in reducing the inventories. The principles of lean management call for minimum inventories (Sawhney et al. 2009), and they should be reasonably followed in the entire value chain.

The redefinition of current, outdated payment terms could also provide a possibility to release working capital in the value chains. E-commerce with advance payments becoming more and more common may support the new mindset regarding payment terms. The radical reduction of payment terms would be possible, as current technology—e.g. by taking advantage of RFID technology—would enable even immediate payments right after the customer has received the delivery from the supplier. Electronic invoicing and automatic invoice handling would support this. However, differences in the negotiation powers of companies play a remarkable role in the renegotiation of new payment terms and sets companies in certain positions, even if win-win situations and benefits for all value chain members should be in target.

4.5   Concept for monitoring and managing the working capital of the value chain

In this chapter, we propose a concept for measuring, monitoring and managing working capital at the value chain level. The proposed concept, WCM platform, is described in Fig. 4.4.

An elementary requirement for the concept is that the value chain has reached certain maturity, which enables collaborative actions for working capital management as well as it enhances openness and trust between the value chain partners.

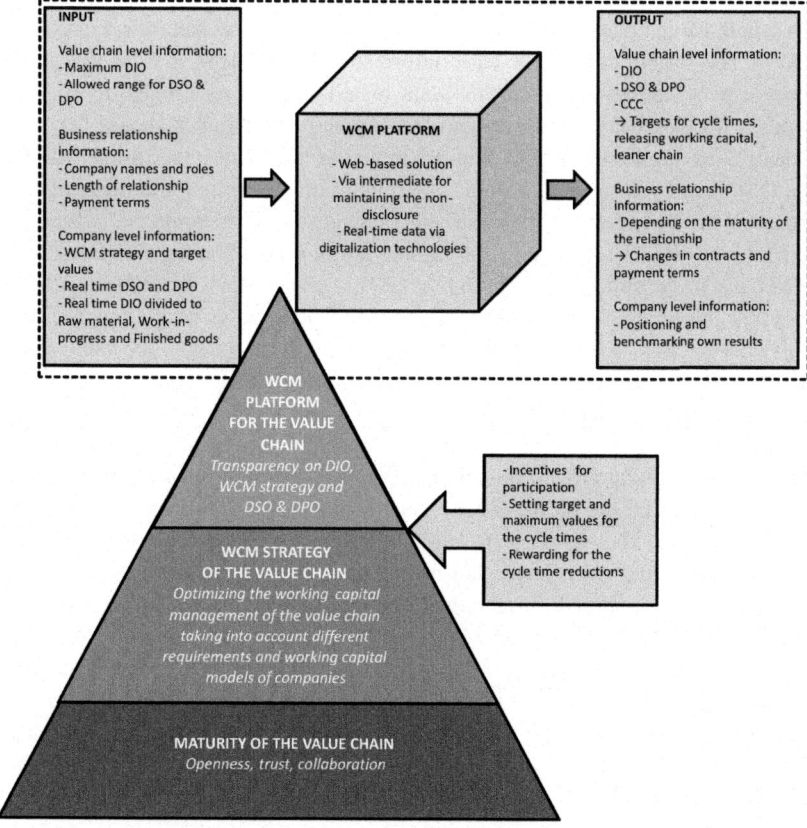

Fig. 4.4   WCM platform

The working capital management strategy for the value chain should be set by considering the different requirements and working capital models of the companies. Target cycle times should be set for both the value chain and individual companies. At the value chain level, the maximum value for the DIO as well as the allowed range for DSO and DPO should be discussed and defined in the WCM strategy. In case of an individual company, the main focus is on the DIO as payment term reductions should be done collectively throughout the value chain.

The main element of the concept is the WCM platform, a web-based tool for measuring, monitoring and managing the working capital of the value chain. Each registered company should provide two types of data: business relationship information and company level information. First, a company should register its business relationships with company names and roles (customer/supplier), and the length of and payment terms for each relationship. Company level data needed for the system consists of basic information on the defined working capital management strategy and real-time data regarding the cycle times of working capital. Working capital management strategy and target values should be given when registering to the system, and updated if needed, but the real-time data on the DSO and DPO, as well as DIO (preferably divided into cycle times of raw material, work-in-progress, and finished goods inventories) should be updated on a regular basis: at least monthly, but in the future, it could be done on weekly or even daily basis with the tools of digitalization. As an output, the WCM platform provides information of the CCC, DIO and financial flows of the value chain. This data can be used in setting new targets for the cycle times in the value chain to release more working capital. The reduction of working capital leads to leaner financial value chains. Information on business relationships can be shared between the business partners and used in negotiations regarding contracts and payment terms. Naturally, information sharing and its use depend on the maturity of the relationship, but on the other hand, using the platform could also enhance the relationship. The WCM platform provides transparency on working capital management in the value chain, but maintains the non-disclosure of the data by operating via an intermediary, and therefore the company names can remain anonymous.

The purpose of the WCM platform is to enable the linking and managing of the physical and financial flows in the value chain via modern technologies and real-time information flow. This is the first proposal of the WCM platform, and therefore some limitations of the platform require further inspection. The reliability of the data has to be secured, and it is needed to ensure that the information flow is consistent with the physical and financial flows. For example, the information flow regarding the quantity and scheduling of the material delivery needs to be equivalent to the real-life physical flow in order to ensure that the basis for the corresponding payment (financial flow) is accurate. Also the commitment of companies is important, and it could be motivated by incentives for participation and by rewards for reductions in the DIO.

4.6   Conclusions and future research

Digitalization offers various possibilities for more efficient working capital management in the value chains. The transparency of cycle times, monitoring working capital in real-time, and linking physical and financial flows in the value chains by taking advantage of modern technologies could release remarkable amounts of working capital to be directed to other investments for strengthening the growth of companies and the value chain. Collaborative working capital management is one important element for gaining competitive advantage in today's business environment and for increasing competition between the value chains instead of individual companies. We proposed a web-based tool, WCM platform, for the collaborative working capital management of the value chain. The WCM platform combines the working capital data of the value chain actors and produces information for monitoring and managing inter-organizational working capital.

The WCM strategy and platform aim at optimizing the working capital management at the value chain level through collaboration between companies. However, more research on defining the optimal working capital management for the value chain and individual companies within it is needed. Also the WCM platform could be developed further to support collaborative working capital management and test it with the pilots before implementation.

References

Ali-Marttila, M., Marttonen-Arola, S., Yli-Kujala, A., Ukko, J., Rantala, T., Sinkkonen, T., Pekkola, S., Saunila, M., Pekkarinen, O., Kärri, T.: Stagewise process towards collaborative and value-driven decisions in maintenance networks. In: Koskinen, K. et al. (eds.) Proceedings of the 10th World Congress on Engineering Asset Management (WCEAM 2015). Lecture Notes in Mechanical Engineering, Springer, Cham (2016)
BMW Group, Annual Report 2009.: https://www.bmwgroup.com/content/dam/bmw-group-websites/bmwgroup_com/ir/downloads/en/2009/BMW_Group_2009_EN.pdf   (2010). Accessed 8 Oct 2010

Deloof, M.: Does working capital management affect profitability of Belgium firms? J. Bus. Finan. Account. **30**(3–4), 573–587 (2003)

Enqvist, J., Graham, M., Nikkinen, J.: The impact of working capital management on firm profitability in different business cycles: evidence from Finland. Res. Int. Bus. Finan. **32**, 356–369 (2014)

García-Teruel, P.J., Martínez-Solano, P.: Effects of working capital management on SME profitability. Int. J. Manag. Finan. **3**(2), 164–177 (2007)

Gehani, R.R.: Time-based management of technology: A taxonomic integration of tactical and strategic roles. Int. J. Oper. Prod. Manag. **15**(2), 19–35 (1995)

Germany Trade & Invest, Industrie 4.0 Smart manufacturing for the future.: https://www.gtai.de/GTAI/Content/EN/Invest/_SharedDocs/Downloads/GTAI/Brochures/Industries/industrie4.0-smart-manufacturing-for-the-future-en.pdf (2014). Accessed 27 Jan 2017

Grosse-Ruyken, P.T., Wagner, S.M., Jönke, R.: What is the right cash conversion cycle for your supply chain? Int. J. Serv. Oper. Manag. **10**(1), 13–29 (2011)

Gupta, S., Dutta, K.: Modeling of financial supply chain. Eur. J. Oper. Res. **211**(1), 47–56 (2011)

Hofmann, E., Kotzab, H.: A supply-chain oriented approach of working capital management. J. Bus. Logist. **31**(2), 305–330 (2010)

Hofmann, E., Maucher, D., Piesker, S., Richter, P.: Ways Out of the Working Capital Trap, Heidelberg (2011)

Huff, J., Rogers, D.S.: Funding the organization through supply chain finance: a longitudinal investigation. Supply Chain Forum: Int. J. **16**(3), 4–17 (2015)

Jose, M.L., Lancaster, C., Stevens, J.L.: Corporate returns and cash conversion cycles. J. Econ. Finan. **20**(1), 33–46 (1996)

Lazaridis, I., Tryfonidis, D.: Relationship between working capital management and profitability of listed companies in the Athens stock exchange. J. Finan. Manag. Anal. **19**(1), 26–35 (2006)

Lind, L., Pirttilä, M., Viskari, S., Schupp, F., Kärri, T.: Working capital management in the automotive industry: Financial value chain analysis. J. Purch. Supply Manag. **18**(2), 92–100 (2012)

Lind, L., Monto, S., Kärri, T., Schupp, F.: Detecting working capital models in the ICT supply chains. Int. J. Supply Chain Invent. Manag. **1**(3), 233–249 (2016)

Lorentz, H., Solakivi, T., Töyli, J., Ojala, L.: Trade credit dynamics during the phases of the business cycle—a value chain perspective. SCM. Int. J. **21**(3), 363–380 (2016)

Monto, S., Lind, L., Kärri, T.: Working capital models: avenues for financial innovations. In: Proceedings of the XXIV ISPIM Conference—Innovating in Global Markets: Challenges for Sustainable Growth 16–19 June 2013, Helsinki, Finland (2013)

Mullins, J.W., Komisar, R.: Getting to plan B: Breaking through to a better business model. Harvard Business Press, Boston, USA (2009)

Ng, B., Ferrin, B.G., Pearson, J.N.: The role of purchasing/transportation in cycletime reduction. Int. J. Oper. Prod. Manag. **17**(6), 574–591 (1997)

Pirttilä, M.: The cycle times of working capital: Financial value chain analysis method. Lappeenranta, Finland (2014)

Pirttilä, M., Viskari, S., Kärri, T.: Working capital in the value chain: cycle times of pulp and paper industry. In: Proceedings of the 19th annual IPSERA Conference: Supply Management—Missing Link in Strategic Management. Lappeenranta, Finland (2010)

Pirttilä, M., Viskari, S., Lind, L., Kärri, T.: Benchmarking working capital management in the inter-organisational context. Int. J. Bus. Innov. Res. **8**(2), 119–136 (2014)

Protopappa-Sieke, M., Seifert, R.W.: Benefits of working capital sharing in supply chains. J. Oper. Res. Soc. (2016)

Richards, V.D., Laughlin, E.J.: A cash conversion cycle approach to liquidity analysis. Finan. Manag. **9**(1), 32–38 (1980)

Seifert, D., Seifert, R.W., Protopappa-Sieke, M.: A review of trade credit literature: opportunities for research in operations. Eur. J. Oper. Res. **231**(2), 245–256 (2013)

Sawhney, R., Kannan, S., Li, X.: Developing a value stream map to evaluate breakdown maintenance operations. Inter. J. Ind. Syst. Eng. **4**(3), 229–240 (2009)

Shin, H., Soenen, L.: Efficiency of working capital management and corporate profitability. Finan. Pract. Educ. **8**(2), 37–45 (1998)

Talha, M., Christopher, S.B., Kamalavalli, A.L.: Sensitivity of profitability to working capital management: A study of Indian corporate hospitals. Int. J. Manag. Finan. Acc. **2**(3), 213–227 (2010)

Talonpoika, A.-M., Monto, S., Pirttilä, M., Kärri, T.: Modifying the cash conversion cycle: revealing concealed advance payments. Int. J. Prod. Perform. Manag. **63**(3), 341–353 (2014)

Valeo.: 2011 Registration Document containing the annual financial report. http://www.valeo.com/medias/upload/2012/12/3095/2011-registration-document-2011-annual-report.pdf (2012). Accessed 14 Oct 2016

Viskari, S., Kärri, T.: A model for working capital management in the inter-organisational context. Int. J. Integr. Supply Manag. 7(1/2/3), pp. 61–79 (2012)

Viskari, S., Kärri, T.: A cycle time model for analysing the efficiency of working capital management in a value chain. Int. J. Bus. Perform. Supply Chain Model. **5**(3), 221–238 (2013)

Viskari, S., Lukkari, E., Kärri, T.: State of working capital management research: Bibliometric study. Middle Eastern Finan. Econ. **5**(14), 99–108 (2011a)

Viskari, S., Pirttilä, M., Kärri, T.: Improving profitability by managing working capital in the value chain of pulp and paper industry. Int. J.Manag. Finan. Account. **3**(4), 348–366 (2011b)

Viskari, S., Lind, L., Kärri, T., Schupp, F.: Using working capital management to improve profitability in the value chain of automotive industry. Int. J. Serv. Oper. Manag. **13**(1), 42–64 (2012a)

Viskari, S., Ruokola, A., Pirttilä, M., Kärri, T.: Advanced model for working capital management: Bridging theory and practice. Int. J. Appl. Manag. Sci. **4**(1), 1–17 (2012b)

Wuttke, D.A., Blome, C., Henke, M.: Focusing the financial flows of supply chains: an empirical investigation of financial supply chain management. Int. J. Prod. Econ. **145**(2), 773–789 (2013)

About the Authors

Lotta Lind is a Junior Researcher at the School of Business and Management, Lappeenranta University of Technology, Finland. After receiving her MSc (Tech.) in 2011, she has worked as a researcher in a working capital management project as well as in the private sector. She is currently on study leave from her job at KONE, a global company in the elevator and escalator business, and is finalizing her PhD studies and dissertation on working capital models in the value chains.

Timo Kärri is a Professor at the School of Business and Management, Lappeenranta University of Technology, Finland. He received his DSc (Tech.) in Industrial Engineering and Management, and his dissertation considered the timing of capacity changes in capital intensive industries. Kärri's current research topics include industrial asset management, capital investments and working capital management, and he has specialized in life-cycle costing and cost modeling. With his research group C^{3}M, he has written over 90 publications, including 32 scientific journal articles. Kärri has extensive teaching experience in the above mentioned areas. He has also acted as the responsible manager in many industrial research projects.

Sari Monto (née Viskari) received her PhD in 2013. After an academic career in the Lappeenranta University of Technology, Finland, she has worked as a Controller for Stora Enso, a global provider of renewable solutions in packaging, biomaterials, wooden construction, and paper. In her research, the main focus area was inter-organizational working capital management.

Miia Pirttilä is a postdoctoral researcher at the School of Business and Management, Lappeenranta University of Technology, Finland. She received her PhD in 2014, and her dissertation focused on the cycle times of working capital in the value chain context. Her research interests include capital, capacity and cost management.

Dr.-Ing. Florian Schupp is Senior Vice President Purchasing Automotive and Automotive Aftermarket of Schaeffler Group. He completed his PhD at the Technical University of Berlin in the field of strategy development in purchasing and logistics and has 19 years of purchasing experience in the companies Schaeffler, Continental, Siemens and SONY. Dr.-Ing. Schupp integrates practical purchasing and supply management work with academic research together with the University of Lappeenranta, Finland, and the University of Catania, Italy. He teaches International Procurement at the Technical University of Berlin and Purchasing and Supply Management at Jacobs University in Bremen. Dr.-Ing. Schupp works in the research fields purchasing strategy, behavioral aspects in purchasing, integration of supplier innovation, parametric purchasing auctions, working capital management, buyer-supplier relations and supply management. In 2014, Dr.-Ing. Schupp was appointed to join the Advisory Board of BLG Logistics Bremen, Germany.

Digitalisierung und Krise

5

Dirk Adam, Florian Glunz und Klaus Kost

Zusammenfassung

Die Digitalisierung wird die Gesellschaft und die Wirtschaft in den nächsten Jahren umfassend beeinflussen. Aus Sicht der Verfasser ist dabei insbesondere zu berücksichtigen, dass mit dem Fortschreiten der Digitalisierung auch die potenzielle Gefahr ökonomischer Krisen oder sogar Insolvenzen steigt. Es soll anhand von Thesen aufgezeigt werden, welche perspektivischen Chancen und Herausforderungen aus ökonomischer und auch juristischer Betrachtungsweise mit der Digitalisierung verknüpft sein können.

5.1 Einleitung

Unternehmenskrisen und Insolvenzen führen oftmals zu einem Abbau von Arbeitsplätzen. Besonders im Falle der Stilllegung und Liquidation eines Unternehmens fällt ein großer Teil der Arbeitsplätze ersatzlos weg. Dies hat neben

D. Adam (✉)
Heidelberg, Deutschland
E-Mail: dirk.adam@wellensiek.de

F. Glunz · K. Kost
Essen, Deutschland
E-Mail: florian.glunz@pcg-projectconsult.de

K. Kost
E-Mail: klaus.kost@pcg-projectconsult.de

© Springer Fachmedien Wiesbaden GmbH 2018
F. Schupp und H. Wöhner (Hrsg.), *Digitalisierung im Einkauf,*
https://doi.org/10.1007/978-3-658-16909-1_5

individuellen Nachteilen für die betroffenen Arbeitnehmer auch erhebliche volks-
wirtschaftliche Konsequenzen zur Folge. So beliefen sich die unmittelbaren Schäden der Insolvenzen in Form von Forderungsausfällen der betroffenen Gläubiger,
bezogen auf Deutschland, im Jahr 2014 auf ca. 26,1 Mrd. EUR (Creditreform
2014, S. 6). Daneben sind nicht genau bezifferbare, aber gleichwohl erhebliche
indirekte Schäden in Folge des Arbeitsplatzwegfalls zu konstatieren. Zusätzlich
kommt es zu Auftragsrückgängen bei Lieferanten, erhöhten Einkaufskosten ehe-
maliger Kunden, Einnahmeausfällen der Finanz- und Sozialkassen sowie Kosten
für Lohnersatzleistungen. Dass die Digitalisierung potenziell eine neue Welle an
Insolvenzen auslösen kann, soll im Folgenden dargestellt werden.

Laut einer Studie des Global Center For Digital Business Transformation (2015,
S. 3 ff.; 18), in welcher 941 Geschäftsführer aus zwölf Branchenzweigen der gan-
zen Welt zum Thema Digitalisierung befragt worden sind, werden rund 40 % der
befragten Unternehmen aus den unterschiedlichsten Gründen durch die Digitalisierung in den kommenden Jahren negativ beeinflusst oder gar vom Markt ver-
drängt. Eine der wesentlichen Folgen der Digitalisierung: Risikofreudige und
innovative Unternehmen, die zudem stellenweise keine eigenen physischen Werte
besitzen, treten innerhalb kürzester Zeit auf den Markt und verändern bzw. verdrängen bestehende Marktstrukturen. Als Beispiele für diese sogenannten disrup-
tiven Veränderungen lassen sich u. a. der Dienstleister Uber oder die Gesamtheit an
„Shared-Mobility"-Dienstleistern anführen. Dabei wird das Wachstum der „Shared-
Mobility" wesentlich durch das Carsharing vorangetrieben (Roland Berger 2014,
S. 10). Laut einer Studie von Roland Berger (2014, S. 10) wird der Carsharing-
Markt bis 2020 ein jährliches Umsatzwachstum von 30 % erfahren.

Aus diesen Ausführungen lässt sich Folgendes ableiten: Bedingt durch das
große Thema Digitalisierung werden Unternehmen in Deutschland in der unmittelbaren Zukunft vermehrt in Krisen und im schlimmsten Fall sogar Insolvenzen
geraten. Bisweilen fehlt im wissenschaftlichen Kontext die thematische Verzahnung von Digitalisierung und Krise bzw. Insolvenz. Einzig die Studie von Struktur Management Partner (2016), in welcher die Auswirkungen der Digitalisierung
auf Automobilzulieferer näher analysiert werden, bildet hier eine Ausnahme. Das
weitgehende Fehlen einer wissenschaftlichen Diskussion zum Thema Digitalisierung in Verbindung mit unternehmerischer Krise oder im gesteigerten Fall sogar
Insolvenz hat die Autoren dieses Beitrags dazu bewegt, einen Aufsatz hierzu bei-
zusteuern.

Für eine bessere wissenschaftliche Durchdringung der Thematik haben die
Verfasser acht Thesen aufgestellt, an welchen die Trends, Entwicklungen und
Risiken der Digitalisierung in Bezug auf Krisen und Insolvenzen verdeutlicht

werden. Im Kontext des Buchtitels sei an dieser Stelle angemerkt, dass der Einkauf innerhalb dieses Kapitels nicht ausschließlich in den Fokus der Betrachtung genommen worden ist. Dass aber auch der Einkauf der Verkettung von Digitalisierung und Krise und daraus drohender Insolvenz nicht entgehen kann, haben Studien unlängst aufgezeigt. Diese legen nahe, dass die bislang bekannten Organisationsstrukturen bezüglich des Einkaufs zunehmend verschwimmen und sich zugunsten von mehr Transparenz und Zusammenarbeit entlang der Wertschöpfungskette verändern werden (AT Kearney 2014, S. 15).

5.2 These 1: Zulieferer und Händler mit analogem Direktvertrieb werden verdrängt

Als analogen Direktvertrieb verstehen die Autoren sozusagen das Gegenstück zum Vertrieb über das Internet bzw. Vertrieb, der durch das Internet unterstützt wird. Als Beispiel für den analogen Direktvertrieb lässt sich der heute weitverbreitete Vertrieb mittels Außendienstmitarbeitern, die die jeweiligen Produkte für ein Unternehmen anbieten, nennen.

In Zeiten der Digitalisierung, in der u. a. Prozessschritte durch sogenannte cyber-physische Systeme (CPS) verbunden werden oder „die Verfügbarkeit aller relevanten Informationen in Echtzeit" geleistet werden soll (Bundesministerium für Wirtschaft und Energie 2015), kann eine solche analoge Vertriebsform kaum noch als zeitgemäß und konkurrenzfähig angesehen werden. Insgesamt müssen CPS als Verbindung der realen (physischen) Objekte und der informationsverarbeitenden (virtuellen) Objekte verstanden werden (IKT.NRW 2013, S. 4).

Für sich genommen stellt dies zwar noch kein Argument für einen digitalen Direktvertrieb dar, doch lässt sich diese Auffassung stärken durch die Hinzuziehung einer von PwC (2014a, S. 3 ff.) durchgeführten Befragung von über 100 Chefeinkäufern. Einkäufer, die sich sozusagen auf der Gegenseite des Zulieferers oder Händlers befinden, geben an, dass der Bereich des Einkaufs zukünftig wesentlich durch Aspekte der Datenerfassung und -analyse beeinflusst werde. Nur damit lässt sich die heute zunehmend erwartete unternehmerische Verantwortung für alle Stufen der Wertschöpfungskette (und somit auch für den Zulieferer oder Händler) realisieren (AT Kearney 2014, S. 21).

Darüber hinaus bietet der digitale Direktvertrieb für den Käufer (und somit auch den Endverbraucher) den Vorteil, die Preise jederzeit einsehen zu können. Diese Preistransparenz, welche in der jüngsten Vergangenheit einen immer größeren Stellenwert eingenommen hat, gibt den Käufern die Möglichkeit, die Preise intensiv vergleichen zu können (Hecking 2016). Wer sich dieser Transparenz und den damit verbundenen Aspekten wie abnehmender Kundenloyalität sowie steigendem Kostendruck widersetzt, wird nach Einschätzung von Branchenexperten in der nahen Zukunft erhebliche Probleme haben (ebd.). Dass die Digitalisierung vor Branchen, Produkten oder Geschäftsbereichen, die bisweilen nicht „durch den Draht" gebeamt werden können „wie Bankdienstleistungen und E-Bücher" (Reinhold Würth, Stiftungsaufsichtsratsvorsitzender der Würth-Gruppe, zitiert nach Hecking 2016), keineswegs stoppt, lässt sich am Beispiel von Uber aufzeigen. Selbst Experten haben vor zehn Jahren nicht voraussehen können, dass sich das Taxiwesen digitalisieren lässt. Dass auch der Handel mit sogenannten C-Teilen – ein Segment, in welchem zum Beispiel die Würth-Gruppe wichtige Umsätze für sich erzielt – durch die Digitalisierung zukünftig verbessert werden wird, zeigen aktuelle Lösungen in der Demofabrik 4.0 auf dem Campus der RWTH Aachen.

5.3   These 2: Klassische große Industriebetriebe werden nicht pleitegehen

Diese These beruht auf einer nüchternen Zahlenanalyse. Mit zunehmender Betriebsgröße nimmt die absolute Anzahl der Unternehmensinsolvenzen erheblich ab. Verdeutlichen lässt sich dies an den Zahlen für 2016. Circa 95 % aller Unternehmensinsolvenzen betrafen Unternehmen, die zwischen einem und 20 Mitarbeitern beschäftigen (Verband der Vereine Creditreform e. V. 2017). Wird dieser Gedanke um den Aspekt der Digitalisierung erweitert, eröffnet sich ein weiteres Problem zuungunsten von kleinen Betrieben.

Um die betriebliche Implementierung der Digitalisierung zu gewährleisten, ist es seitens der Unternehmen nötig, dass umfangreiche Investitionen getätigt werden. Laut einer Einschätzung der von PwC (2014b, S. 7) veröffentlichten Studie zu dem Thema wird insbesondere im sekundären Sektor ein Investitionsvolumen von betriebsübergreifend 40 Mrd. EUR jährlich bis zum Jahr 2020 notwendig sein, um die Implementierung zu ermöglichen (ebd.). Auch McKinsey (2015, S. 14) betont den enormen Investitionsbedarf durch die Aussage, dass im Zuge von Digitalisierung rund um Industrie 4.0 in naher Zukunft 40 bis 50 % der bestehenden Produktionsanlagen in der Industrie ausgetauscht werden müssen. Werden diese Investitionen nicht getätigt, droht ein erheblicher Wettbewerbsnachteil, welcher neben der Krisen- und Insolvenzgefahr auf Unternehmensebene auch für den Standort Deutschland negative Auswirkungen haben wird. So droht den acht wichtigsten Branchen Deutschlands (darunter die Maschinen- und Anlagenbauindustrie und die Automobilindustrie) ein Verlust an Wertschöpfungspotenzial in Höhe von 220 Mrd. EUR, wenn die Digitalisierung nicht vollumfänglich implementiert wird (Berger 2015, S. 10).

In Anbetracht der absoluten Anzahl von Unternehmensinsolvenzen und dem als enorm zu bezeichnenden Investitionsvolumen für die nahe Zukunft gilt es für kleine- und mittelständische Unternehmen, sich aktiv um die Kooperation mit anderen Unternehmen zu bemühen. In Kooperationsnetzwerken – auch und insbesondere mit großen Unternehmen – lassen sich Synergien heben und Investitionen sinnvoll ergänzend tätigen. Andernfalls wird die Digitalisierung viele kleine und mittelständische Unternehmen in eine Krise führen, die sie nicht erfolgreich bewältigen können. Am Rande sei dazu angemerkt, dass für den letztgenannten Aspekt auch die Beratungsunternehmen und Insolvenzverwalter sich zukünftig noch verstärkter mit den individuellen Problemen, Chancen und Risiken von Digitalisierung auseinandersetzen müssen, um für den Fall der Fälle konkrete Hilfe leisten zu können.

5.4   These 3: Auswirkung auf Logistik – Logistiker müssen sich digitalisieren und in cyber-physische Systeme einbinden

Im Kern ist diese These mit den Aussagen der ersten These vergleichbar, jedoch wird der Fokus hier auf eine weitere Branche gerichtet, welche sehr deutlich und zeitnah die Auswirkungen der Digitalisierung spüren wird.

Eines der zentralen Ziele der Digitalisierung in der Wirtschaft ist die Erreichung der Losgröße 1 (Denner 2014). Dies bedeutet konkret, dass Kundenwünsche individueller berücksichtigt werden können bzw. die Produktion flexibler und reaktionsschneller wird (ten Hompel und Henke 2014, S. 621). Damit die Produktion die nötige Flexibilität und Reaktionsschnelligkeit erreichen kann, muss „die ‚bewegende Instanz‘ der Wirtschaft", die Logistik, diese Ansprüche ebenfalls erfüllen (ten Hompel und Henke 2014, S. 617). Hierfür ist die Verbindung von Produktion und Logistik durch cyber-physische Systeme unerlässlich. Übertragen auf die Wirtschaft bedeutet dies, dass ein vom Kunden online erteilter Auftrag, im Idealfall ohne menschliche Weiterverarbeitung, direkt die Wertschöpfungskette auslöst. Damit einerseits für die Produktion die notwendigen Materialen, Hilfs- und Betriebsstoffe etc. vorhanden sind, andererseits das fertige Produkt – am besten ohne kostenaufwendige Zwischenlagerung „just-in-time" – ausgeliefert werden kann, ist insbesondere die Einbindung der Logistik in CPS von elementarer Bedeutung.

Dies kann allerdings nur geleistet werden, wenn die Logistikunternehmen die Chance gewährt bekommen, die Daten ihrer Kunden zu erhalten, in der Lage sind, diese korrekt zu analysieren, und somit den Weg zur Logistik 4.0 (von Wehberg 2016, S. 324) mit bestreiten können. Wird dieser Gedanke weitergeführt, so hat die Logistik mit der Digitalisierung die nahezu historische Möglichkeit, „das ‚Diktat des Standorts‘ für die Verfügbarkeit von Produkten […] durch sinkende […] Transaktionskosten, steigende Transparenz von Angeboten und Überall-Erhältlichkeit (Ubiquität) von Gütern" aufzuheben (von Wehberg 2016, S. 335). Es wird somit insgesamt sehr deutlich, dass die Logistik einerseits, wie auch zu Beginn dieser These angedeutet, enorme Veränderungen erfahren wird, aber andererseits auch enormes Potenzial zur Veränderung von Wirtschaftsstrukturen und Gesellschaft mit sich bringt.

5.5   These 4: 3-D-Drucker machen den Einkauf überflüssig

Bereits die vorherige These hat in Auszügen angedeutet, dass den vorherrschenden Wirtschaftsstrukturen Veränderungen bevorstehen werden. Wird jedoch der 3-D-Druck in den Fokus der Betrachtung gerückt, so sind die Auswirkungen noch verheerender. Der Sportartikelhersteller Adidas hat unlängst vorgemacht, was durch den 3-D-Druck möglich ist. Durch die sogenannte „Speedfactory" stellt das DAX-Unternehmen in einem flexiblen, kundenorientierten und dezentralen Fertigungsprozess Schuhe innerhalb weniger Stunden durch 3-D-Druck her (Adidas Group o. J.). Bemerkenswert sind hieran zwei verschiedene Aspekte. Zunächst werden 3-D-Drucker zukünftig genutzt, um ein Massenprodukt zu individualisieren, wodurch eines, wenn nicht gar das Ziel von Industrie 4.0 konkret erreicht wird: die Losgröße 1 (vgl. These 3). Andererseits wird eine Produktions(-rück) verlagerung nach Deutschland ermöglicht. Zwar wird hierdurch ein Teil der Wertschöpfung zurück nach Deutschland geholt, wobei anzumerken ist, dass die bei einer Produktion in Asien üblichen 300 Mitarbeiter (entlang der gesamten Wertschöpfungskette bei Schuhen) durch den 3-D-Druck nicht vollumfänglich rückverlagert werden (Köhn 2016) – während der 3-D-Drucker für Deutschland insgesamt als Chance zu verstehen ist (mehr Arbeitsplätze im Anlagenbau und im Bereich der IT), bedeutet er aber auch, dass „China will have to give up on being the mass-manufacturing powerhouse of the world" (D'Aveni 2013), da plötzlich 3-D-Drucker zur Konkurrenz für die Serienproduktion werden (Pfeifer und Niemann o. J.). Somit hat der 3-D-Druck für die weltwirtschaftspolitische Ordnung den bereits angesprochenen verheerenden Charakter.

Doch auch für Deutschland kann der 3-D-Druck negative Effekte haben. So wird z. B. die ohnehin schon als volatil zu bezeichnende Stahlbranche diese Effekte in naher Zukunft wohl sehr deutlich erfahren. In Deutschland hängen nach Angaben des Rheinisch Westfälischen Wirtschaftsinstituts (2015) rund 3,5 Mio. Arbeitsplätze direkt oder indirekt vom Stahl ab. Ein nicht unwesentlicher Anteil dieser Arbeitsplätze entfällt auf den Einkauf bzw. die Distribution von Stahl.

Sollte das Pilotprojekt in Amsterdam, Stahl an Ort und Stelle zu einer Brücke zu gießen, erfolgreich sein, so könnte hierdurch der erste Beweis erbracht sein, dass der Stahlbranche u. a. ein massiver Arbeitsplatzverlust im Einkauf bevorsteht. Es sollte jedoch der argumentativen Vollständigkeit halber darauf hingewiesen sein, dass unabhängig von der Stahlbranche, der Einkauf nicht komplett überflüssig sein wird. Auch für den 3-D-Druck wird ein „klassischer" Einkauf, wenn auch wahrscheinlich in geringerem Umfang, notwendig sein.

5.6 These 5: Insolvenzrestrukturierungen – mehr qualitative Restrukturierung als quantitatives Cost Cutting

Die bisher vorgestellten Thesen thematisieren alle das Themenfeld Digitalisierung als Veränderungselement für bestehende wirtschaftliche Strukturen, mit erheblichen Konsequenzen für Gesellschaft, Staat und Privatleben. Die nun folgenden Ausführungen zur fünften These sollen dabei aufzeigen, welche Probleme in Verbindung mit der Digitalisierung auftreten können, wenn die Zeichen der Zeit nicht erkannt worden sind.

Einer der wichtigsten Gründe für Insolvenzen sind fehlerhafte oder gar gänzlich ausbleibende Investitionen, z. B. in Technologien, wie sie in den vorherigen Thesen dargelegt worden sind (Staab 2015, S. 113). Doch was tun, wenn die Krise oder die Insolvenzsituation bereits da ist?

Die Aufgabe für Geschäftsführungen, Insolvenzverwalter oder Berater ist in solchen Zeiten zunächst trivial anmutend: Kostenreduzierung. Staab (2015, S. 159) bezeichnet, dieses in der Beraterbranche auch als „Cost Cutting" bekannte Prinzip als eines, bei dem Kosten „über ganze Abteilungen verteilt, nach dem ‚Rasenmäherprinzip' gleichmäßig verringert werden sollen." Diesem Ansatz geht jedoch in vielen Fällen eine langfristige bzw. nachhaltige Neuausrichtung und somit Stabilisierung eines Krisenbetriebs nahezu vollkommen ab. Die Alternative ist der Schritt in die entgegengesetzte Richtung. Investitionen in neue Produkte, Prozesse oder Dienstleistungen sind notwendig, um auf diese

Weise den Umsatz und im besten Fall den Ertrag zu steigern (ebd.). Doch Investitionen in innovative Prozesse und/oder Produkte sind unbestreitbar mit hohen Kapitalbindungen und folglich mit hohen Liquiditäts- und Erfolgsrisiken verbunden (ebd. S. 115). Um diese Risiken zu minimieren, darf in einem solchen Prozess der Mitarbeiter als eine der wichtigsten Stellschrauben nicht vergessen werden. In der Beratungspraxis wird vielfach deutlich, dass Mitarbeiter ein detailliertes Problemverständnis innerhalb eines Unternehmens haben und somit auch konkrete Verbesserungs- und Innovationsideen mitbringen. Um dies im Sinne einer qualitativen Restrukturierung zu erreichen, ist es wichtig, dass Rahmenbedingungen geschaffen werden, die dies zulassen. Hierunter fällt zu aller erst eine funktionierende Kommunikationsbereitschaft über alle Hierarchieebenen hinweg, aber auch, für den Fall einer Implementierung aufwendiger Innovationen, die entsprechende Verbesserung der Arbeitnehmerqualifikationen (Disselkamp 2012, S. 89). Es liegt aber auf der Hand, dass sich dies im Verlauf einer Krise, die sich eventuell bis zur Insolvenz steigert, in der die Zeit oft ein limitierender Faktor ist, oftmals schwer umsetzen lässt, womit der dringliche Apell verbunden ist, dies bereits in wirtschaftlich positiveren Zeit zu ermöglichen.

5.7 These 6: Die Umgestaltung der Betriebsorganisation als Voraussetzung für „Einkauf 4.0" – krisenauslösend oder verschärfend?

Die Einführung und Vollziehung von „Einkauf 4.0" erfordert das Umdenken des Marktes und des Managements. Begreift man, dass in dem Schlagwort „Einkauf 4.0" nicht nur ein technisch abstrakter Begriff formuliert ist, sondern zugleich auch eine völlig andere, vom Unternehmer bzw. dem Management zu ändernde Betriebsorganisation angesprochen ist, so wird einem bewusst, dass zur praktischen Umsetzung eines „digitalisierten Einkaufs" eine Strukturveränderung innerhalb der bestehenden Betriebe und letztlich eine erforderliche Änderung der Organisationsstruktur funktionierender Betriebsteile, Geschäftsabläufe und Hierarchien erforderlich wird. So werden die Einkaufsabteilungen, die Logistikabteilungen, die IT-Abteilung mit Anbindung an die Buchhaltung und das Controlling und letztlich auch die Berichtswege innerhalb des Managements grundlegend umorganisiert werden, und es werden neue Strukturen geschaffen werden müssen, um die Betriebe zukunftsfähig zu machen und den künftig an erhöhten Prämissen der Digitalisierung genügenden Markterfordernissen entsprechend anzupassen.

Jeder Restrukturierer weiß, dass die Umgestaltung einer bestehenden und an und für sich funktionierenden Betriebsorganisation ohne konkreten Krisenbezug und „nur" zur Schaffung einer zukunftsträchtigen Formation mit zu den schwierigsten Herausforderungen des Managements der Neuzeit zu zählen ist. Die arbeitsrechtliche Gestaltung einer zukunftsträchtigen Betriebsorganisation ohne eigentlichen wirtschaftlichen, im „Hier und Jetzt" verankerten und objektiv-faktisch nachvollziehbaren Anlass ist in der praktischen Realität ein geradezu selbstzerstörerisches Verlangen, dessen Akzeptanz zumeist nicht rechtzeitig innerhalb der betroffenen Belegschaft erreicht werden kann, weil dort Besitzstandsinteressen – verständlicherweise – vorherrschen bzw. diesen Priorität eingeräumt wird. Hinzu kommt, dass zum Zeitpunkt der beabsichtigten Umstrukturierung die Arbeit für verschiedene Abteilungen (z. B. Einkauf/Vertrieb in alter, überholter Formation und Struktur) noch nicht entfallen ist und die „Ideal-Struktur", die künftigen Anforderungen genügen soll/kann, erst noch aufgebaut werden muss, was eine visionäre Perspektive erforderlich macht, die typischerweise weder von Arbeitnehmern, Betriebsräten, mitbestimmenden Aufsichtsräten/Beiräten oder Tarifvertragsparteien entsprechend erkannt und wohlwollend mitgetragen wird.

Dies bedeutet, dass gerade bei größeren Unternehmen mit hohen Personalanteilen im Bereich Logistik, Einkauf/Vertrieb und Controlling ernsthafte Eingriffe in die Betriebsstruktur unternommen werden müssen und diese nur mit erheblichem Widerwillen der beteiligten bzw. betroffenen Arbeitnehmer, Betriebsräte und Tarifvertragsparteien umgesetzt werden können.

So ist jede grundlegende Änderung der Betriebsstruktur zugleich eine Betriebsänderung im Sinne der § 111 ff. BetrVG und erfordert die elementare Einbindung des (typischerweise bestehenden) Betriebsrates und den Abschluss von Interessenausgleichen, wenn (und das ist der unterstellte Regelfall) gewisse Nämlichkeitsschwellen überwunden werden. Dabei bezieht sich die Nämlichkeitsschwelle auf den Prozentsatz von Personalabbau, ab dem ein Interessenausgleich verhandelt werden muss. Die genauen Werte sind dabei dem § 112a BetrVG zu entnehmen.

Dies bedeutet, dass ein stabiles, operativ gesundes Unternehmen durch den perspektivischen, von Visionen getragenen Schritt, sich zukunftsfähig umzugestalten, in eine ernsthafte kollektivarbeitsrechtliche Herausforderung, wenn nicht sogar Krisensituation versetzt werden kann, indem aus diesem Anlass der Betriebsfrieden und das gedeihliche Miteinander – wenn es dies überhaupt gibt – mit dem Betriebsrat und den Tarifvertragsparteien einer ernsten Herausforderung ausgesetzt werden.

In der praktischen Umsetzung führt dies typischerweise zu einem erheblichen Kostenaufwand im Bereich von Beratungs- und Umsetzungskosten, Abfindungsvolumen und durch die damit verbundenen Friktionen und Stimmungsschwankungen mit der Belegschaft zu erheblichen operativen Minderleistungen, die sich durch erhöhte Fehlerquoten und vermehrte Ausschüsse, erhöhte Krankentage und eine insgesamt geringere Betriebsleistung inhaltlich qualifizieren.

Nicht selten führen Umorganisation und Strukturveränderungen eines gesunden Betriebes zum Zwecke der Zukunftssicherung der Arbeitsplätze paradoxerweise dazu, dass hierdurch erst eine vorher nicht einmal im Ansatz erkennbare oder existente Krise ausgelöst wird und sich die wirtschaftliche Verfassung des Unternehmens insgesamt verschlechtert.

Das Arbeitsrecht, und insbesondere das kollektive Arbeitsrecht, unterscheidet nicht, aus welchem Motiv heraus eine Kündigung ausgesprochen wird oder eine Versetzung bzw. sonstige Personalanpassungsmaßnahme angedacht und durchgeführt werden soll. Damit liegt im deutschen Arbeitsrecht selbst der Grund, warum eine Zukunftssicherung und damit die Sicherung der Arbeitsplätze selbst nur schwer und mit erheblichem Aufwand, zeitlichem Versatz und zusätzlichen Kostenpositionen (Abfindungsvolumen sind im angloamerikanischen Rechtskreis weitgehend oder zumindest in deutschen Dimensionen fremd) verbunden ist. Insoweit ist der Gesetzgeber, nicht zuletzt der europarechtliche Leitlinien-Ersteller in Brüssel, gefragt, diesen Wettbewerbsnachteil in Europa in den Blick zu nehmen und einer – auch den Traditionen einer sozialdemokratischen Marktwirtschaft folgend – Angleichung an das weltweite Entwicklungs- und Dynamikniveau zuzuführen.

5.8 These 7: Zeitliche Aspekte in Krisensituationen – Chancen und Risiko

Erkennt man, welchen Aufwand die Einführung von „Einkauf 4.0" in einer umzugestaltenden Betriebsorganisation erforderlich macht, so erschließt sich hieraus unmittelbar, dass der zeitliche Aspekt von entscheidender Bedeutung sein wird. Da der vorliegende Buchbeitrag insbesondere Krisensituationen in den Blick nehmen möchte, seien drei aus Sicht des Restrukturierers/Praktikers typische Krisenbezüge exemplarisch herausgestellt und vor Augen geführt:

5.8.1   Die vorinsolvenzliche Restrukturierung als Krisenauslöser

Wie bereits weiter oben unter These 6 (Abschn. 8.7) dargelegt, ist es möglich, dass durch die Umorganisation eines Betriebes sowohl in operativer als auch sozialpolitischer Hinsicht eine Krise im Unternehmen ausgelöst wird und damit das Unternehmen sich in eine Krisensituation bzw. Restrukturierungsphase versetzt sieht. Gelingt eine rasche Umsetzung der Umorganisation und Neustrukturierung der betrieblichen Abläufe unter Beachtung der Voraussetzungen für „Einkauf 4.0" nicht schnell genug, so ist es nicht unwahrscheinlich, dass hierdurch eine Insolvenzsituation – insbesondere mit den Chancen der Eigenverwaltung zur Vollendung der Umorganisation unter der privilegierenden Bedingung des Insolvenzrecht (s. hierzu Ziffer 3, Abschn. 5.8.3) – bewusst oder unbewusst, beabsichtigt oder unbeabsichtigt verursacht wird. Damit kann die Einführung von „Einkauf 4.0" selbst ein „Brandbeschleuniger" für eine bestehende Krisensituation sein bzw. eine solche Unternehmenskrise erst auslösen.

5.8.2   Die Unmöglichkeit der „Modernisierung zu Einkauf 4.0" im Insolvenzverfahren

Im Insolvenzverfahren selbst ist es dem Insolvenzverwalter typischerweise nicht gestattet, im Rahmen der Betriebsfortführung das Unternehmen grundlegend umzustrukturieren und einen modernen, digitalisierten Einkauf zu etablieren. Zwar ist es dem Insolvenzverwalter gestattet (und es ist sogar seine Pflicht), Betriebsteile, die operativ nicht zu halten sind und daher auch nicht übertragungsfähig sind, stillzulegen und die dort verorteten Arbeitsplätze abzubauen, jedoch wird es unter Kosten-Nutzenrelation (und dies ist der relevante Aspekt aus Sicht der Gläubiger, die dem Insolvenzverwalter harsche Kostendisziplin unter dem ständigen Impetus der Massemehrung aufbürden) nicht zu rechtfertigen sein, den Kostenaufwand eines digitalisierten Einkaufs in der Fortführungsphase eines Unternehmens umzusetzen und damit die Wettbewerbsfähigkeit des Unternehmens wieder herzustellen.

Damit steht als Zwischenergebnis fest, dass im Rahmen eines Insolvenzverfahrens eine Umstrukturierung mit dem Ziel der Etablierung eines „Einkaufs 4.0" üblicherweise nicht umgesetzt werden kann.

5.8.3   Postinsolvenzliche Chancen für Erwerber und Investoren

Ist ein Unternehmen wegen einer rückschrittlichen Betriebsorganisation und Unternehmensstruktur in die Krise geraten, so ist es einem Investor/einem Erwerbervehikel möglich, unter den Prämissen des sogenannten Erwerberkonzeptes Vermögenswerte im Wege des Asset Deals aus der Insolvenzmasse vom Insolvenzverwalter zu erwerben und unter Vorlage eines Erwerberkonzeptes den Insolvenzverwalter zu sogenannten Veräußererkündigungen zu motivieren, die einem privilegierten Kündigungsschutz unterfallen, sodass es dem Erwerber ermöglicht wird, einen Betrieb aus der Insolvenz heraus zu kaufen, der den von ihm gewünschten betrieblichen und organisatorischen Anforderungen entspricht.

Damit kann ein Erwerber z. B. verlangen, dass zum Zwecke der Einführung von „Einkauf 4.0" eine betriebliche Struktur vom Insolvenzverwalter geschaffen wird, die der Erwerber sodann als Ganzes im Wege des Asset Deals und in Vollziehung eines Betriebsübergangs nach § 613 a BGB erwerben kann.

5.8.4   Zwischenfazit

Das Paradoxon des deutschen Restrukturierungs- und Insolvenzrechts wird an dieser Stelle plastisch: Nach geltendem Recht ist es nur schwer möglich, Unternehmen außerhalb der Insolvenz grundlegend umzugestalten und damit eine Krise zu vermeiden oder aufzulösen. Ist hingegen die Krise in eine echte Insolvenzsituation kulminiert, so ist es dem Erwerber (d. h. einem Dritten) aber möglich, diese Struktur unter den Privilegien des reduzierten Kündigungsschutzes (auf Basis eines arbeitsplatzsichernden Erwerberkonzeptes) wettbewerbsfähig und zukunftsfähig zu gestalten, um es sodann als Ganzes an den Erwerber, Investor zu übertragen und die dort belegten Arbeitsplätze zu retten.

5.9   These 8: Europarechtliche Dimension – Krisenforcierung aus Brüssel oder notwendige Incentivierung eines uninspirierten, nicht mehr wettbewerbsfähigen Marktes?

Die Einführung und Fortentwicklung von „Industrie 4.0", insbesondere mit Blick auf den Einkauf, birgt erhebliche rechtliche Gefahren.

Als Basis von „Modernisierung 4.0" kann ein ständiger effektiver Informationsaustausch betrachtet werden. Dieser erfordert die Benutzung bestimmter Plattformen sowie den Zugang zu diesen. Problematisch ist, dass bislang keine expliziten Regelungen darüber bestehen, welche Daten der Anbieter bei Benutzung seiner Plattform einsehen und ggf. weiterleiten oder gar zu eigenen Zwecken nutzen darf. Um an dem Fortschritt mithalten zu können, sind Unternehmen jedoch darauf angewiesen, Zugang zu solchen Plattformen zu erhalten. Gleichermaßen soll eine drohende Monopolisierung – verursacht durch nur wenige vorhandene Anbieter oder ein aufgrund besonders hoher Praktikabilität als Standard sich durchsetzendes Angebot – verhindert werden.

Ein in der Sache vergleichbarer Fall ereignete sich bei einem weltführenden Software- und Hardware-Unternehmen, welches seinen Wettbewerbern den Zugang zu selbstentwickelten Plattformen und Programmen verwehren wollte, um somit die marktbeherrschende Stellung beizubehalten. Das Gericht der Europäischen Union hat dies jedoch verhindert, indem es das Unternehmen dazu verpflichtete, den Zugang zu seinem Betriebssystem zu öffnen.

Ein weiteres Problemfeld bei der Einführung und Fortentwicklung von „Industrie 4.0" stellt der Datenschutz dar. Unterschiedliche Vorschriften im Datenschutzrecht der Mitgliedsstaaten der Europäischen Union behindern den Handel,weswegen bereits seit Längerem eine Harmonisierung der nationalen Vorschriften postuliert wird. Insoweit wurde – in der Vergangenheit – die Datenschutz-Richtlinie 95-46-EG verabschiedet, an deren Regelungsinhalte sich die Vorschriften des Bundesdatenschutzgesetzes und der Landesdatenschutzgesetze anpassen mussten, teilweise aber auch den eingeräumten Spielraum der Datenschutz-Richtlinie über Gebühr ausnutzten und teilweise sogar weit darüber hinausgingen.

Dies hat zu kontroversen Diskussionen mit anderen EU-Mitgliedsstaaten und der Kommission geführt. Eine weitergehende Harmonisierung des Datenschutzrechts als Voraussetzung für den „Einkauf 4.0" ist daher auf allen Ebenen – national und europaweit – gewünscht und an einer entsprechenden Verordnung wurde bereits seit 2012 gearbeitet.

Die Datenschutzgrundverordnung (EU) 2016/679 ist nunmehr bereits in Kraft getreten, aber erst ab Mai 2018 anwendbar. Sie ersetzt grundsätzlich nationale Vorschriften, enthält aber Öffnungsklauseln und Spielräume für Mitgliedsstaaten, die diese in nationales Recht umsetzen müssen, hier konkret durch Änderung des Bundesdatenschutzgesetzes und der Landesdatenschutzgesetze.

Ein Problem ist die Anwendbarkeit dieser Datenschutzrichtlinie auf internationale Sachverhalte und besteht in den hierbei auftretenden Regelungslücken. So ist die Datenschutzrichtlinie beispielsweise nicht bei Verarbeitung von Daten im Inland durch Stellen mit Sitz in der EU, aber ohne Niederlassung in der EU anwendbar[1].

Ein grundlegendes, zeitnah zu lösendes Problem besteht darin, dass im Bereich des Datenschutzes durch Verlegung von Niederlassungen ins nichteuropäische Ausland bewusst die Anwendungsbereiche europarechtlicher Richtlinien umgangen werden können. Gerade bei Logistik-Unternehmen und exportstarken Unternehmen, bei denen „Einkauf 4.0" von herausgehobener Bedeutung sein dürfte, kann durch einfache Niederlassungsverlagerung ein Wettbewerbsvorteil dadurch entstehen, dass man den strengeren Datenschutzrichtlinien der EU nicht mehr Folge leisten muss. Hierdurch wird der Industriespionage Tür und Tor geöffnet. Aufgabe der EU ist es jedoch, gerade eine solche Ausspähung zum Schutz des unternehmerischen Eigentums zu verhindern.

Insoweit mag insbesondere die Europäische Union auch den hiesigen Kontinent überlagernde Sachverhalte in den Blick nehmen und die weltweit zur Anwendung gelangenden Datenschutzrichtlinien, die miteinander im Wettbewerb stehen und wichtige Standortvorteile für die Zukunft definieren, bestmöglich zu harmonisieren versuchen.

Auf EU-Ebene werden bereits weitere Maßnahmen gegen Cyber-Angriffe angestrebt. Die Kommission beabsichtigt, in der noch beratenen NIS-Richtlinie (Richtlinie zur Netz- und Informationssicherheit) einheitliche Mindeststandards einzuführen. In Einklang mit der Richtlinie soll jeder Mitgliedstaat eine hohe Netz- und Informationssicherheit schaffen. Die NIS-Richtlinie soll ferner dazu dienen, aktuell bestehende Regelungslücken im europäischen und nationalen Datenschutzrecht zu schließen.

Auch Unternehmen sollen dieses Ziel vor Augen haben und entsprechende geeignete und angemessene Maßnahmen ergreifen.

Für die Einführung und Fortentwicklung von „Industrie 4.0" ist ein allumfassendes Datenschutzrecht unerlässlich. Damit ein effektiver Datenaustausch gewährleistet wird, muss eine ganzheitliche Fortentwicklung im europäischen Rahmen erfolgen. Es müssen Vorkehrungen auf datenschutzrechtlicher sowie auf

---

[1]Ein US-amerikanischer Anbieter betreibt ein soziales Netzwerk auf Deutsch und hat eine Niederlassung in Deutschland, die Werbeflächen auf dieser verkauft und sich u. a. dadurch finanziert. Nutzer schließen Vertrag mit irischer Niederlassung, die die personenbezogenen Daten der Nutzer im Rahmen einer Auftragsverarbeitung an den Mutterkonzern in die USA weiterleitet.

wettbewerbsrechtlicher Ebene getroffen werden. Bei konkreten technischen Ent-wicklungen – autonomes Fahren oder elektronische Deichsel – kann die Auto-matisierung und Datennutzung auch nur zu einer Anpassung der Gesetzeslage in einzelnen Bereichen (Straßenverkehrsvorschriften und Haftungsbestimmungen) führen.

Es bleibt gespannt abzuwarten, ob insbesondere die Europäische Union sich dieser Aufgabe gewachsen sieht und es schafft, technischen Fortschritt und damit einhergehende Risiken angemessen in Einklang zu bringen.

5.10 Fazit

Die Zukunft hält Herausforderungen für die Wirtschaft bereit, die – wie es schon Charles Darwin erkannt hat – zu einer Selektion und damit zu einer Verfeinerung der im Wettbewerb stehenden Unternehmen führen wird. In den Thesen 1 bis 4 wurde dargestellt, wie die Autoren sich abzeichnende branchentypische Entwick-lungen einschätzen. Wie üblich bei Thesen haben auch diese keinen Anspruch auf konkrete Verwirklichung in der Zukunft. Viel mehr sind die Thesen als Bei-spiele für die wirtschaftlichen Potenziale der Digitalisierung zu verstehen und sollen einen Appell an alle Entscheidungsträger richten: Die Digitalisierung muss jetzt mit allen Facetten gestaltet werden. Hierbei, und das soll These 5 in aller Deutlichkeit aufzeigen, dürfen die Mitarbeiter in keinem Fall vergessen werden. Gleichermaßen stellt die These 5, aber auch die folgenden Thesen 6 und 7, den vorstellbaren Krisenverlauf samt seiner Herausforderungen, Probleme und Chan-cen im Insolvenzfall dar. Darauf aufbauend wurde in These 8 das Pflichtenheft der Politik in persona eines noch nicht allseits akzeptierten „europäischen Gesetz-gebers" angesprochen. Allen Einzelthesen ist gemein, dass in der nicht allzu fernen Zukunft ein erheblicher Änderungs- und Anpassungsbedarf auf die Wett-bewerbsteilnehmer der Marktwirtschaft zukommt. Dieser Abschnitt hat mit dem Verweis auf Darwin begonnen und es soll an dieser Stelle mit dem wohl bekann-testen Zitat von ihm enden. Die Digitalisierung, so umfangreich und folgenreich diese jetzt schon ist und in Zukunft noch werden wird, wird zu einem „survival of the fittest" führen.

Literatur

Adidas Group (Hrsg.): Adidas errichtet erste SPEEDFACTORY in Deutschland. Herzogen-aurach. http://www.adidas-group.com/de/medien/newsarchiv/pressemitteilungen/2015/adidas-errichtet-erste-speedfactory-deutschland/ (o. J.). Zugegriffen: 20. Febr. 2017

AT Kearney (Hrsg.): Procurement 2020+ – 10 Mega-Trends, die den Einkauf verändern werden. O. O. (2014)

Berger, R. (Hrsg.): Think Act. Shared Mobility – How new businesses are rewriting the rules of the private transportation game. Roland Berger Strategy Consultants GmbH, München (2014)

Berger, R. (Hrsg.): Die digitale Transformation der Industrie. Springer, Berlin (2015)

Bundesministerium für Wirtschaft und Energie (Hrsg.): Die vierte industrielle Revolution gemeinsam gestalten (inkl. Unterseiten). Berlin. http://www.plattform-i40.de (2015). Zugegriffen: 3. Jan. 2017

Creditreform (Hrsg.): Insolvenzen in Deutschland. Neuss. https://www.creditreform.de/fileadmin/user_upload/crefo/download_de/news_termine/wirtschaftsforschung/insolvenzen-deutschland/Analyse_Insolvenzen_in_Deutschland__Jahr_2014.pdf (2014). Zugegriffen 17. Okt. 2017

D'Aveni, Richard A.: 3-D Printing will change the world. Harv. Bus. Rev. Innov. 2013(3) (2013)

Disselkamp, M.: Innovationsmanagement. Instrumente und Methoden zur Umsetzung in Unternehmen, 2. Aufl. Springer, Wiesbaden (2012)

Denner, V.: Der Schlüssel zum Erfolg. Handelsblatt 12.12.2014 (2014)

Global Center For Digital Business Transformation (Hrsg.): Digital Vortex – How digital disruption is redefining industries. IMD, Lausanne (2015)

Hecking, M.: Kampf um Würths Erbe: Wie der Onlinehandel das B2B-Geschäft umwälzt. http://www.manager-magazin.de/unternehmen/handel/kampf-um-wuerths-erbe-wie-der-onlinehandel-das-b2b-geschaeft-umwaelzt-a-1112263.html (2016). Zugegriffen: 20. Febr. 2017

IKT.NRW (Hrsg.): Industrie 4.0 Cyber Physical Systems. Nordrhein-Westfalen auf dem Weg zum digitalen Industrieland. Wuppertal. https://cps-hub-nrw.de/file/2800/download?token=6VguPSZM (2013). Zugegriffen: 17. Okt. 2017

Köhn, R.: Adidas: Schuhe aus der Hochgeschwindigkeitsfabrik. http://www.faz.net/aktuell/wirtschaft/internet-in-der-industrie/adidas-schuhe-aus-der-hochgeschwindigkeitsfabrik-14251157.html (2016). Zugegriffen: 20. Febr. 2017

McKinsey (Hrsg.): Industry 4.0 How to navigate digitization of the manufacturing sector. O. O. (2015)

Pfeifer, M., Niemann, P.: Wie der 3D-Druck Märkte verändert. http://www.digital-engineering-magazin.de/fachartikel/wie-der-3d-druck-maerkte-veraendert (o. J.). Zugegriffen: 31. Jan. 2017

PWC (Hrsg.): Einkauf – Die neue Macht in den Unternehmen. PWC, Frankfurt (2014a)

PWC (Hrsg.): Industrie 4.0 – Chancen und Herausforderungen der vierten industriellen Revolution. PWC, Frankfurt a. M. (2014b)

Staab, J.: Die 7 häufigsten Insolvenzgründe erkennen und vermeiden. Wie KMU nachhaltig erfolgreich bleiben. Springer, Wiesbaden (2015)

Ten Hompel, M., Henke, M.: Logistik 4.0. In: Bauernhansel, T., ten Hompel, M., Vogel-Heuser, B. (Hrsg.) Industrie 4.0 in der Produktion, Automatisierung und Logistik, S. 615–624. Springer, Wiesbaden (2014)

Verband der Vereine Creditreform e. V. (Hrsg.): Unternehmensinsolvenzen 2016: Kleinstunternehmen markieren Trend (inkl. Unterseiten). http://www.creditreform.de/nc/aktuelles/news-list/details/news-detail/unternehmensinsolvenzen-2016-kleinstunternehmen-markieren-trend-2907.html (2017). Zugegriffen: 23. Jan. 2017

Wehberg, G.G.von: Logistik 4.0 – die sechs Säulen der Logistik in der Zukunft. In: Göpfert, I. (Hrsg.) Logistik der Zukunft – Logistics for the Future, 7. Aufl, S. 319–344. Springer, Wiesbaden (2016)

Über die Autoren

Dirk Adam ist Rechtsanwalt und Partner bei WELLEN-SIEK in Heidelberg. Mit seinem generalistischen Ansatz berät er ganzheitlich in Krisensituationen und Insolvenzverfahren. Nach dem Abschluss seiner juristischen Ausbildung trat Dirk Adam 2004 als Rechtsanwalt bei WELLENSIEK ein und wurde 2010 zum Partner ernannt. Vor dem Hintergrund seiner Expertise im allgemeinen und krisennahen bzw. insolvenzbezogenen Prozessrecht, im Arbeitsrecht und auch im Wirtschafts-, Insolvenz- und Steuerstrafrecht übernimmt er neben mittelständischen Mandaten mitunter auch ausgelagerte Tätigkeiten für Insolvenzverwalter der Sozietät.

Florian Glunz ist Berater bei der PCG – Project Consult GmbH in Essen, einer Unternehmensberatung mit dem Fokus auf Arbeitnehmer und Mitbestimmung. Seit dem Abschluss seines Masterstudiums der Wirtschaftsgeographie im Jahr 2016 ist Florian Glunz für die PCG – Project Consult in der Beratung u. a. zu den Themen Restrukturierung und Digitalisierung tätig. Aufgrund seines Studiums und seiner früheren Tätigkeit als wissenschaftlicher Assistent am Lehrstuhl „Stadt- und Regionalökonomie" an der Ruhr-Universität Bochum ist er im Unternehmen ebenfalls für die Anfertigung von Studien und Gutachten mitverantwortlich.

Prof. Dr. Klaus Kost ist studierter Wirtschaftsgeograph. Nach seiner zehnjährigen Tätigkeit als Lehrbeauftragter an der Ruhr-Universität in Bochum (1987–1997), wurde Klaus Kost dort 1998 Professor am Geografischen Institut. Seit 1998 ist er außerdem geschäftsführender Gesellschafter der PCG – Project Consult GmbH in Essen, einer arbeitsorientierten Beratungsgesellschaft in den Handlungsfeldern Unternehmenssanierung, Beschäftigungssicherung und Innovationsförderung.

Autonomous Manufacturing-related Procurement in the Era of Industry 4.0

6

Yilmaz Uygun und Maria Ilie

Abstract

The Fourth Industrial Revolution or Industry 4.0 will create intelligent manufacturing systems that consist of highly autonomous and self-controlled machines, equipment, and material that communicate with each other not only within a company but also across companies in order to organize the value chain and produce the intended products in the best possible way with less or even no human interaction. In this context, procurement becomes an interesting issue that needs to be considered in detail in such manufacturing environments. This chapter analyzes the effect of those intelligent manufacturing systems on procurement processes by distinguishing between procurement tasks that will likely stay in human hand and those that will be taken over by those intelligent machines and systems. Especially for the latter scenario, a concept for an operational, fully-automated, and manufacturing-related procurement system is developed and presented here. For this, we begin by looking at the manufacturing-related procurement types in the era of Industry 4.0 that may be done by machines, followed by the discussion of innovation-based procurement practices that will require human input. We conclude by presenting the general concept of the autonomous manufacturing-based procurement system. Although the concept for the autonomous manufacturing-based

Y. Uygun (✉) · M. Ilie
Bremen, Deutschland
E-Mail: y.uygun@jacobs-university.de

M. Ilie
E-Mail: m.ilie@jacobs-university.de

© Springer Fachmedien Wiesbaden GmbH 2018
F. Schupp und H. Wöhner (Hrsg.), *Digitalisierung im Einkauf,*
https://doi.org/10.1007/978-3-658-16909-1_6

procurement system is fully developed, further research is needed to connect this system to intelligent machines to enable a fully autonomous procurement especially to execute transactional relations and repetitive tasks.

6.1   Introduction

In the era of Industry 4.0, machines and other equipment will be highly intelligent and make their own decisions not only concerning in-company processes but also those crossing company boarders in terms of procurement. The latter is true for machines that can independently communicate to other machines and management information systems in the same as well as other. Those machines will have the freedom of choice for repetitive and even more complex decisions in terms of scheduling and procurement. Smaller order and manufacturing lot sizes paired with short-lived demand—that will even become more challenging in the era of Industry 4.0—will lead to high scheduling efforts and also to scheduling conflicts that require a permanent rescheduling, both is supposed to be performed by those intelligent machines.

Due to those changing requirements, in cases of external procurement the search for a capable supplier needs to be sped up. This requires also an adequate level of data release in a central database or cloud so that a quick procurement and rescheduling can take place.

Currently there is no IT support for such a case. On the one hand there are technical means to support company-wide scheduling but they are not accessible to external companies or even machines and were not designed for such cases. On the other hand, some web-based solutions are available which act as listings and do not represent integrated solutions. Thus, an instrument is necessary which provides standardized decision rules and procedures for short-term manufacturing-related procurements.

Nevertheless, some procurement decisions will most likely still remain in human hand, such as strategic and long-term procurement decision and especially innovation-related decisions when it comes to new product developments. These kind of decisions require emotional intelligence and advanced skills and knowledge in human behavior and market opportunities that may not be the strength of machines.

So, this chapter deals with manufacturing-related procurement issues in the era of Industry 4.0 by looking at innovation-related procurement needs on the one hand and a concept for an operational, fully-automated, and manufacturing-related procurement system.

6.2   Literature Review on Procurement 4.0

The interface of Industry 4.0 and procurement is a rather novel topic that will definitely be more taken into consideration in the near future. Looking at current literature in this realm, it becomes evident that many papers, concepts, and applications revolve around e-procurement.

Harrigan et al. (2008) agree that, while there are many benefits to e-procurement in manufacturing, the prerequisite of reaping the said benefits is the correct implementation of e-procurement. They report that, while e-procurement is employed frequently in manufacturing, its use can be quite conservative, with only basic online catalogues and request for tenders through traditional EDI systems being mentioned by the surveyed companies. The implications of this finding are that the potential synergies between buyers and sellers are not exploited due to suboptimal e-procurement implementation. E-procurement systems, however, could help shift the power from the sellers in favor of the manufacturing companies, should these problems be solved. Company size also plays a part in how successful the manufacturers can be in implementing e-procurement due to the strategic nature of the investment.

However, not all manufacturers use e-procurement conservatively. Davila (2003) also observes a category of aggressive adopters of e-procurement that pave the way for more conservative companies. While this overall trend shows signs of slowing down, it is still crucial in increasing adoption of e-procurement across the industry, which will lead to many synergies once implemented extensively.

Apart from that, there are also some contributions that focus on public e-procurement in emerging countries and its efficiency. Gurakar and Tas (2016) analyze the case of Turkey and conclude that there are negative effects to introducing e-procurement in public auctions due to the current barriers to e-procurement, which should be eliminated by policymakers in order to reap the whole benefits of the process. Several other studies perform similar case studies, such as Janevski et al. (2016) for Macedonia and Panduranga (2016) for India, that agree upon the fact that e-procurement is beneficial when implemented correctly.

The most relevant contribution for this chapter is by Glas and Kleemann (2016). They highlight the importance of distinguishing between e-procurement and Industry 4.0, as the concepts are intertwined and many scholars argue that their delineation might not constitute a new development. The authors go on by calling e-procurement the next development step after MRP and ERP, but before Procurement 4.0. The literature is unclear whether there should be a distinction between Industry 4.0 and Procurement 4.0.

So, it becomes apparent that there is less discussion on the interface of industry 4.0 and procurement, let alone concepts and techniques that support the digitalization. This chapter addresses this issue and presents a concept on the basis of Uygun (2012) that contributes to the digitalization of manufactured-related procurement processes in an intelligent and self-controlled manufacturing environment.

6.3 Manufacturing and Related Procurement in the era of Industry 4.0

6.3.1 Manufacturing in the era of Industry 4.0

Before delving deeper into procurement issues once the fourth industrial revolution has taken place, the concept of manufacturing in the era of Industry 4.0 and related key terms need to be defined. For this, Fig. 6.1 gives an overview of the relation of those key terms in a nutshell.

The Internet of Things (IoT) is a global infrastructure enabling advanced services by interconnecting (physical and virtual) things based on existing and evolving interoperable information and communication technologies (ITU 2012). A

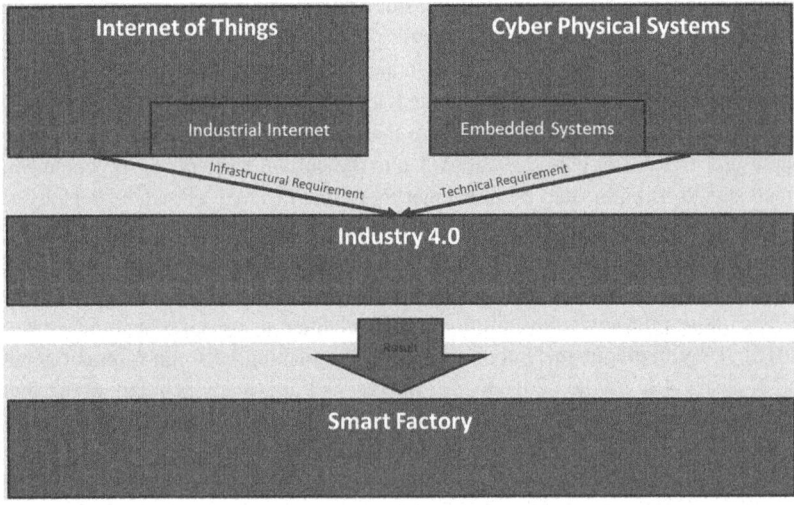

Fig. 6.1 Key terms in advanced manufacturing. *Source* Uygun and Reynolds (2017)

special manifestation of this is the Industrial Internet that is an Internet of Things, machines, computers, and people, enabling intelligent industrial operations using advanced data analytics for transformational business outcomes (IIC 2015).

In addition to that, the term cyber-physical systems (CPS) is frequently used in this realm. CPS are embedded systems that are connected to each other using digital networks and have several multi-modal human-machine interfaces and gather and process physical data through sensors and interact with physical operations through actors (acatech 2011). The core of such CPS are embedded systems which are microprocessor-based systems that are built to control a (range of) function and are not designed to be programmed by the end user (Heath 2003).

Both IoT and CPS are the infrastructural and technical requirements respectively for the proclaimed fourth industrial revolution or Industry 4.0 (CPG 2013).

The outcome of Industry 4.0 is smart factories that have a decentralized manufacturing logic through intelligent products in horizontally and vertically integrated production systems for a consistent engineering along the value chain (CPG 2013). Figure 6.2 depicts such a smart factory.

So, how does manufacturing and related procurement work in such smart factories? Orders may be placed conventionally, by intelligent machines or machine parts

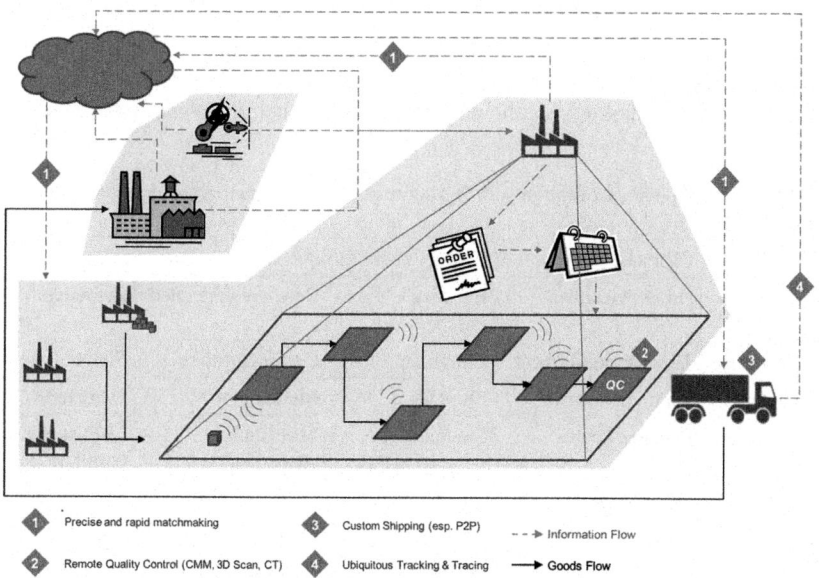

Fig. 6.2 An ideal manufacturing environment in Industry 4.0

(e.g. in the case of wear and tear or failure), or out of the cloud. The latter refers to the scenario where machines or management information systems seek suppliers immediately for a part or component. Those orders are then scheduled autonomously. Based on the order load in terms of quantity, type, and time, the appropriateness of the current manufacturing layout will be reassessed and adjusted accordingly in a fully automated manner. Accordingly, the raw material that is equipped with sensors and actors will find its way through the shop floor by communicating with the machines and other equipment, such as forklifts, etc. The finished goods may then be inspected for quality assurance purposes by individual and specialized service providers that offer their services in the cloud just like the delivery service providers. The delivery of finished goods to those quality service providers and to customers is handled by individual resource-sharing delivery service providers.

In such a scenario, that does not consider innovation-related procurement efforts, procurement encompasses machine parts, material, quality, and delivery services alongside conventional raw material procurement that may show different characteristics which is discussed in the following section.

6.3.2 Manufacturing-based Procurement Types

Based on the future scenario depicted in Fig. 6.2 different order types may occur. Those types differ in several characteristics. To better understand and classify them, a morphology, as shown in Table 6.1, is helpful. First and foremost,

Table 6.1 Morphology of the manufacturing-related procurement types. *Source* based on Uygun (2012)

Category	Characteristic	Characteristic Attributes		
Initiation	Order placement	Reactive unscheduled	Reactive scheduled	Proactive
Capability	Type of requirement	Congruent	Complementary	Substituting
Scheduling	Planning horizon	Short term	Medium term	Long term
	Planning stages	Single stage	Multistage sequential	Multistage parallel
Product	Complexity	Standard parts	Complex parts	
	Provision of material	With provision	No provision	
Process	Process complexity	Standard process	Special process	
	Number of processes	One process	Several processes	

procurement-related categories, such as initiation, capability, scheduling, product, and process, can be distinguished that are logically connected to each other.

The morphology in Table 6.1 allows the determination and specification of four typical categories of automated and self-controlled procurement relations that may occur in the mentioned manufacturing systems 4.0 of the future. These are the

- problem-based procurement
- capability-based procurement
- technology-based procurement
- blanket-order procurement

The problem-based procurement is relevant in case of process-related issues, such as break downs, failures, etc. The affected process to produce the corresponding parts will be kept in-house. Thus, just a temporary by-pass is installed. The need here is triggered reactively unscheduled due to the sudden occurrence of a problem. Congruent or similar capabilities are required. Due to the sudden occurrence of the problem, a short-term single-stage adaptation planning is required. Primarily, standard parts are purchased, whereas semi-finished parts or raw material are supplied as input. This process has a low complexity, which allows for a transactional procurement.

The capacity-based procurement can technically be seen as a capacity adjustment. On the basis of existing orders the capacity requirement is determined, which necessitates a capacity-related procurement, if the internal capacity level is exceeded. The planning of the capacity-based procurement is thus triggered reactively. Here, depending on the problem, either complementary/supplementary or congruent/similar capabilities are considered. The planning horizon is days to weeks or months, with single to multi-stage planning. The products may be standard or complex that may or may not require raw material supply. Thus, the outsourced processes can be seen as standard or special in terms of technology. One or more processes are outsourced here.

If processes or operations of a particular technology are outsourced, a technology-related procurement is considered. Here, economic considerations play a role. The in-house processing of these processes is unfavorable in terms of costs that make those processes eligible to be outsourced to specialized external manufacturing service providers, e.g. coating. The corresponding capabilities are not kept internally (anymore). The demand is determined in advance. Depending on the company's situation with the presence of own capabilities, the capability requirements can be complementary or substituting. In the former case, there are no own capabilities, whereas in the latter case own capabilities are given up. Since this kind of subcontracting must be well thought through, the planning horizon may take several months. The planning can include one to several steps.

The blanket-order procurement is based on predefined and already negotiated framework contracts that allow demand-oriented ad-hoc procurements on short notice. The order placement is mainly reactive scheduled for complementary capabilities. Dependent on the negotiations, the planning horizon may be medium to long-term for multi-stage and rather complex and numerous processes that generate complex parts and do not need any provision of material by the customer.

The latter procurement type requires more personal interaction than the others since it is based on pre-negotiated contracts. In addition to that, anything else that goes beyond those operative and manufacturing-related procurement types, such as innovation-related procurement efforts, which are discussed in the following section, need intense personal interaction.

6.4   Innovation-based Procurement Practices

Even in the era of Industry 4.0 with highly autonomous and intelligent machines and equipment, innovation-related decision as to new product development and introduction are supposed to be a human task. Our recent research (IPC 2015; Reynolds and Uygun 2017) shows that over the past five to ten years, many Original Equipment Manufacturers (OEM) have undergone a significant reorganization and rethinking of their supply chains in terms of innovation and procurement. Pressures, primarily financial, from customers have forced them to rethink how best to drive greater efficiency and innovation from the supply chain. A few of the relevant changes that have taken place include:

- An integration of supply chain management with engineering to bring design and technological innovation into the supply chain procurement process earlier.
- Centralizing supply chain operations across business units or particular products rather than within each business unit.
- A consolidation of the supply chain to reduce the overall number of suppliers and hence complexity.
- Greater emphasis on collaborative partnerships with a select number of strategic suppliers, and a more solutions-oriented approach with suppliers in general.

- Shortening lead times overall and highly responsive supply chains to respond to customer demands that cannot be known ahead of time.
- Increasing globalization of the supply chain so that suppliers can be sourced from any corner of the globe as long as they are cost competitive and deliver quality products on time.

These changes directly impact suppliers within supply chains. OEMs generally want to build a strong supplier base that is reliable and can work with them long term. OEMs regard their suppliers as crucial to their own business success, as ever more activity has been outsourced. They further emphasize that they seek deep, strategic relations rather than transactional relations with their key suppliers, i.e. those making hard-to-source, mission-critical technology or components. Those strategic suppliers of OEMs are the top of a supplier pyramid representing the stratification of suppliers according to their value-added as it relates to innovation. The vast majority of the supplier base can be regarded as commodity and bottleneck suppliers who are providing less critical parts and components than those who are seen as strategic suppliers.

The actual number of suppliers who are considered truly critical or strategic suppliers is small; only 10–15% of OEMs' supply base (composed of around a thousand firms), and sometimes considerably less. Those suppliers provide key technology where labor is not the key cost driver and price not the primary consideration in sourcing.

These changes have significant implications for suppliers with respect to what it takes to be a top and reliable supplier. Irrespective of industry, OEMs today have similar expectations of their suppliers. Criteria, such as standard certifications, technical skills, high quality standards, on-time delivery, cost reductions, are seen as standard requirements for their top suppliers.

OEMs are conscious that suppliers need to receive sufficient margin to enable reinvestments, and that starving suppliers of this is ultimately self-defeating. At the same time, regular price reductions are a basic expectation of suppliers by OEMs. This is true because driving annual cost reductions is a basic metric procurement managers are assessed upon. Many of the OEMs are themselves exposed to fierce price competition. Also, supplier annual cost reductions are a leading indicator of that firm's ability to engage in process improvements and remain competitive.

OEMs generally believe that the way out of this contradiction between ensuring sufficient margins for suppliers and achieving the expected price/cost reductions, is thorough adoption of lean practices—not only within the supplier firms but, ideally, throughout the supply chain, to make the entire system lean (and not just the individual nodes of the chain).

Several OEMs singled out the importance of deep collaboration between engineering teams from the supplier, the OEM and the OEM's customer in the design of the final products and of the components going into them, as one of the surest ways to sustainable cost reductions.

In practice, while most OEMs agree that ideally supply chain relations should be structured as deeply cooperative partnerships, they also note that at present these relations do not exceed more than a handful of suppliers. OEMs expect from new suppliers innovative components that both add value to OEMs' products and support OEMs' product innovation process.

On the other hand, OEMs have to commit themselves to particular aspects to ensure a sustainable and long-lasting relationship through communication of business roadmaps, long-term contracts, direct and indirect financial support for their suppliers, and business opportunities for (inter-)national expansion.

Clearly, the relationship between OEMs and SMEs is evolving toward higher standards but also more collaboration. Many OEMs are taking the "high road" in terms of investing in their suppliers to help them be more productive and potentially grow. In response to these significant changes within their supply chain, many OEMs have created supplier development initiatives for those suppliers with whom they want to develop a long-term relationship, like e.g. seconding engineers or other technical staff helping suppliers manage their processes and supply chain, guaranteeing work and/or helping broker credit lines to allow suppliers to purchase new equipment, and supporting suppliers in following the OEM abroad.

So, it becomes apparent that those strategic aspects of procurement will remain and even become more important in the era of Industry 4.0. As discussed earlier, transactional relations with a limited or even high number of options may be automated more easily in the era of industry 4.0. However, it is hard to imagine that highly intelligent robots, machines, and other equipment can come up with creative ideas that lead to innovations for human beings which simultaneously can evoke strong emotions at the consumers/customers.

6.5 Manufacturing-related Operational Procurement 4.0

Based on the above-mentioned manufacturing-related procurement types that are eligible for autonomous procurement by intelligent machines in the era of Industry 4.0, a general concept of an autonomous manufacturing-based procurement system with its technical description is presented in this section.

6.5.1   General Concept of the Autonomous Manufacturing-based Procurement System

A machine in a manufacturing system in the era of Industry 4.0 that experiences scheduling conflicts may trigger the procurement by means of a software client (see Fig. 6.3). Subsequently, the corresponding requirements are specified.

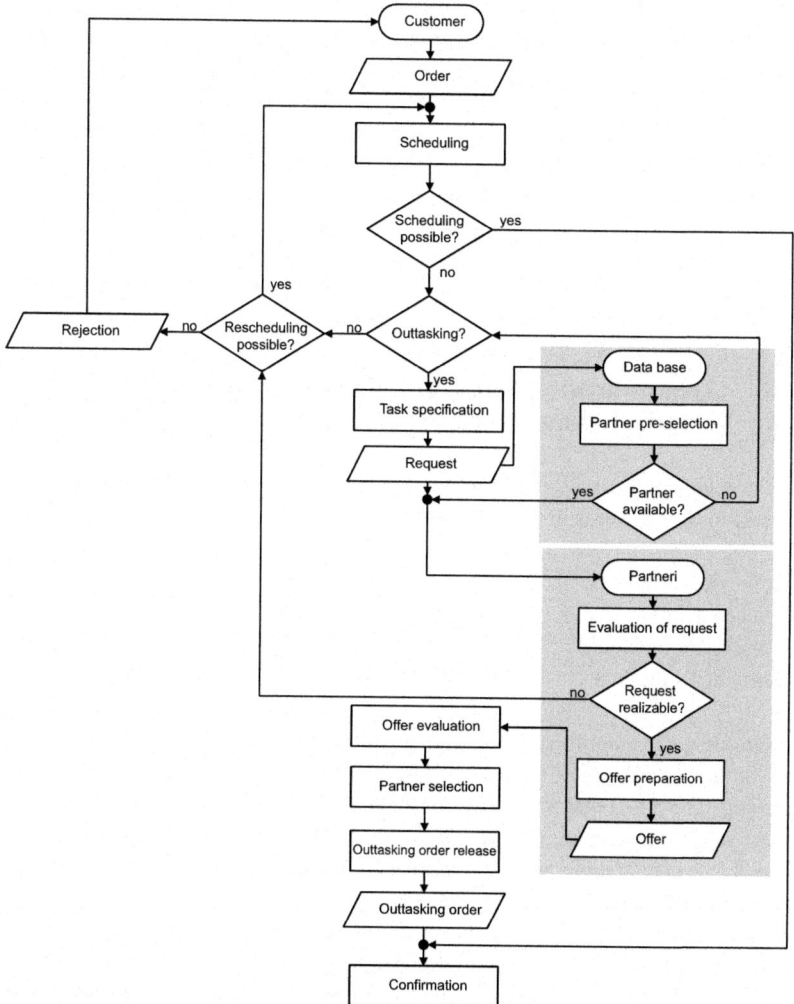

Fig. 6.3 The specific process model. *Source* Uygun (2012)

Thereafter, the request is sent automatically. For this, the database/cloud in which all participating companies are registered is scanned for potential partners, i.e. companies and their respective machines, first. Potential partners are those that could handle the required operation due to the possession of certain capabilities.

After retrieving information from the cloud, the potential suppliers (represented by a business software or even machines) are contacted automatically by the customer's software client to get their time-wise availability. Those suppliers then respond through their software clients whereas that time slot is then blocked on the affected machines. Only those suppliers that signal a time-wise availability will in the next step get all necessary order specification documents (technical drawings etc.) to check whether they have the technical skills to fulfill that request. The suppliers can do this through automatically scanning the drawings and pattern recognition of price categories in historical order data to calculate and submit qualified values based on similarity profiles. Accordingly, the customer's software client evaluates all submitted offers and selects a partner whom he would like to award the contract to. Finally, this job will be scheduled by both partners.

6.5.2 Communication Structure of the Autonomous Manufacturing-based Procurement System

The communication structure of this system follows the general procurement processes according to (Presutti 2003), such as

- Define requirements,
- Select Supplier,
- Contract Agreement,
- Supplier Evaluation.

The autonomous manufacturing-based procurement system combines independent, geographically distributed participants with no central control. For this, a technology is needed that is suitable for the realization of the distributed structure of such kind of procurement systems. In recent years the multi-agent technology emerged in which actually existing entities are represented as software agents for complex situations that handle certain tasks autonomously.

So, the basic structure of the autonomous manufacturing-based procurement system takes this technology into consideration and includes a database/cloud and a software client that operate in the background by communicating with each other through software agents.

The software client is used for inter-company/inter-machine communication in the case of capability search and is linked to the database/cloud. The appropriate procurement will be selected and specified, the search for potential partners will be started, and finally the order will be submitted. Suppliers submit their offers and price bids by means of their software client.

The development of the communication structure of the multi-agent system, which uses the above-described components of the system, is performed on the basis of the PASSI procedure model according to Cossentino et al. (2005), ICAR/CNR (2008), and Chella et al. (2006). For this, the phases of the functional system description, agent identification, role identification, task identification, ontology description, role description, multi-agent implementation, and implementation of individual agents are run and completed with the application configuration.

In the first phase, the functional description of the total system is performed based on the so-called use case diagrams. In preparing these and other diagrams, the UML standard is used.

Agents who are behind the identified functions can be grouped together. This is necessary for the phase of the agent identification. Here, the functionality of each agent is defined. An agent includes one or more individual functions. Agents can exchange messages with each other, send important information to place an order, add a machine to the database/cloud, match the company's login information with the database, and look for potential companies that have a required machine.

The agents ClientAgentLogIn, ClientAgentMachineAdd, ClientAgentJob and ClientAgentMachineDelete are relevant for both the supplier and customer. The ClientAgentJob collects all offers of service providing companies. For the customer the ManagerAgentCustomer is a central agent that is started automatically. After this, it hooks up to the database first and then to companies/machines that have the desired machine tool or competence. Henceforth, it is the central contact partner for communication with the appropriate agent of potential partner companies by exchanging information with the intermediary agents ClientAgentJob and DBAgentJob. It sends the job data directly to potential suppliers and receives offers from those through intermediary agents.

For the supplier/service provider the ManagerAgentProvider is the counterpart of the ManagerAgentCustomer. The ManagerAgentProvider maintains contact with the appropriate agent of the customer and the intermediary agents, by visualizing the requests, submitting offers, transferring the job to scheduling, and sending the final confirmation. In the database, there exist both the DBAgentMachineListener that selects potential companies with the right machine from the database and the DBAgentJob. The latter is a mediator between the manager agents of the customer and provider. It collects the offers and the price bids of each manager agent of potential partner companies and forwards it to the manager agent of the customer. For the sake of completeness, there are also further database agents, like DBAgentLogin, DBAgentMaschineAdd, and DBAgentMachineDelete, which interact with the aforementioned manager agents.

The next phase, namely the role identification, is based on the identification of all the possible scenarios of the actions of the identified agents to fulfill their function in the context of the multi-agent system. These are illustrated by the well-known UML sequence diagrams in which the roles of each agent can be seen. The exchanged messages in the sequence diagrams refer to the internal communication between the agents. A message in this case defines the executed role of an agent. The data of such a message are specified in the following ontology description phase.

Hereafter, the phase of the task specification follows, in which the individual behavior and tasks of the agents are presented based on activity diagrams. A task includes certain functionalities that form a logical unit of work. A transition between the various activities may refer to the communication between agents, but also to a transition to a new state.

The semantics of the multi-agent system is described in terms of explicit ontologies for which class diagrams are used. On the one hand, the domain ontology, in which the existing entities are represented by classes, is considered. The communication ontology with the agent knowledge and the communication relationship of the agents is addresses on the other hand.

The domain ontology is described as an XML scheme. It represents a class diagram describing the different classes of the multi-agent system. The domain is represented using concepts, actions, predicates, and relationships between them. The various classes may consist of simple data types, such as integer or string. With the help of the class diagrams, the ontologies of the agents are defined, described and shown. First, the ontology of the domain is described, showing the components involved in the software by individual classes, such as order, machine, company, customer, provider, offer, and database. The elements in this diagram define the agent knowledge and the ontology of communication.

That communication ontology is then used for the description of the agent knowledge and communication, where each agent is represented by a single class. This includes his knowledge. The agent communication is specified regarding ontology elements, language, and protocols.

In the role description the life cycle of an agent with its roles, his need for cooperation, and its communication is modeled. The individual roles of the agents in the society and the tasks for each role are detailed. Here, individual roles are represented by classes. Accordingly, the class diagram for such presentation is applicable. This class diagram is extended by the attributes of the classes in the next step, which results in the multi-agent structure.

Within the multi-agent implementation, the structure and the behavior of the agents are defined, which influence each other. It represents a model of the system architecture based on classes with the appropriate attributes and methods and includes the structure definition and the definition of behavior.

Within the single agent implementation, the structure and behavior of each agent are defined, whereas the structural and behavioral definitions are also distinguished here. For each agent, a class is defined by which the inner structure of the agent is determined with its main and sub-classes necessary for the task identification.

In the last phase of the use configuration, the positions of the agents in the distributed system are defined. There is a distinction between database and users, the latter are further subdivided in customer and provider. In the database all corresponding database agents are accommodated which interact with the agents of the users. User data, machine data as well as order data are exchanged.

Eventually, it has to be stated that each company has both the ManagerAgentCustomer and the ManagerAgentProvider, since a company can function both as capacity demander and capacity provider.

Based on this concept, which builds upon Uygun (2012), an automated communication in cases of the aforementioned manufacturing-related procurement gets possible.

6.6   Conclusion and Outlook

Summing up, in the era of Industry 4.0 with highly intelligent and autonomous machines and equipment transactional relations are likely to be executed by those entities. The presented concept in this chapter shows how such automated transactions can be designed in terms of the technical specifications. With this concept, machines may independently communicate to other machines and management information systems in other companies. Those machines will have the freedom of choice for repetitive and even more complex decisions in terms of scheduling and procurement. The shift in order management toward smaller lot sizes and short-lived demand will lead to high scheduling and rescheduling efforts due to scheduling conflicts that can easily be handled by those intelligent machines using such kind of autonomous manufacturing-based procurement systems.

However, innovation-based procurement processes that require a higher degree of creativity and strategic thinking are likely to be still performed by humans. So, even in an intelligently automated manufacturing environment human procurement officers and managers will be highly valuable.

Although the concept is fully developed, further research is needed to connect this system to machines to enable a fully autonomous procurement for cases discussed in this chapter.

References

acatech: Cyber-Physical Systems—Innovationsmotor für Mobilität, Gesundheit, Energie und Produktion (acatech POSITION), vol. 11. Springer, Heidelberg (2011)

Chella, A., Cossentino, M., Sabatucci, L., Seidita, V.: Agile PASSI: an agile process for designing agents. Int. J. Comput. Syst. Sci. Eng. **21**(2), 133–144 (2006)

Cossentino, M., Gaglio, S., Sabatucci, L., Seidita, V.: The PASSI and agile PASSI MAS meta-models compared with a unifying proposal. In: Pechoucek, M., Petta, P., Varga, L. (Eds.) Multi-agent systems and applications IV—4th International Central and Eastern European Conference on Multi-Agent Systems—Proceedings, pp. 183–192. Springer, Berlin (2005)

CPG—Communication Promoters Group of the Industry-Science Research Alliance & National Academy of Science and Engineering: Recommendations for implementing the strategic initiative Industrie 4.0—Final report of the Industrie 4.0 Working Group, Frankfurt a. M. (2013)

Davila, A., Gupta, M., Palmer, R.: Moving procurement systems to the internet: the adoption and use of e-procurement technology models. Eur. Manage. J. **21**(1), 11 (2003)

Glas, A. H., Kleemann, F. C.: The impact of industry 4.0 on procurement and supply management: A conceptual and qualitative analysis. Int. J. Bus. Manag. Invent. **5**(6), 55–66 (2016)

Gurakar, E.C., Tas, B.O.: Does public e-procurement deliver what it promises? Empirical evidence from Turkey. Emerg. Markets Finance Trade **52**(11), 2669–2684 (2016)

Harrigan, P.O., Boyd, M.M., Ramsey, E., Ibbotson, P., Bright, M.: The development of e-procurement within the ICT manufacturing industry in Ireland. Manage. Decis. **46**(3), 481–500 (2008)

Heath, S.: Embedded systems design. Newnes, Oxford (2003)

ICAR/CNR: PASSI. http://www.pa.icar.cnr.it/passi/index.html (2008). Accessed 12 Dec 2016

IIC—Industrial Internet Consortium: What is the industrial internet? http://www.industrial-internetconsortium.org/about-industrial-internet.htm (2015). Accessed 12 Dec 2016

IPC—Industrial Performance Center: Strengthening the innovation ecosystem for advanced manufacturing—pathways and opportunities for Massachusetts. Industrial performance center. Massachusetts Institute of Technology (2015)

ITU—International Telecommunication Union: Recommendation ITU-T Y.2060—Overview of the internet of things. http://www.itu.int/ITU-T/recommendations/rec.aspx?rec=y.2060 (2012). Accessed 12 Dec 2016

Janevski, Z., Davitkovska, E., Petkovski, V.: The constraints of SMEs in the republic of Macedonia in the process of public e-procurement. Econ. Dev./Ekonomiski Razvoj **18**(1/2), 179–193 (2016)

Panduranga, V.: Transparency in public procurement through e-procurement in India. J. Internet Banking Commer. **21**(3), 1–7 (2016)

Presutti Jr., W.D.: Supply management and e-procurement: creating value added in the supply chain. Ind. Mark. Manage. **32**(3), 219–226 (2003)

Reynolds, E. B., Uygun, Y.: Strengthening Advanced manufacturing innovation ecosystems: The case of Massachusetts. Technological Forecasting and Social Change. https://doi.org/10.1016/j.techfore.2017.06.003 (2017)

Uygun, Y.: Integrierte Kapazitätsbörse. Dissertation, Verlag Praxiswissen, Dortmund (2012)

Uygun, Y., Reynolds, E.B.: Advanced manufacturing ecosystems. In: Jeschke, S., Brecher, C., Song, H., Rawat, D.B. (Eds.) Industrial internet of things—cybermanufacturing systems, pp. 691–715. Springer, Berlin (2017)

About the Authors

Prof. Dr. Dr.-Ing. Yilmaz Uygun is Professor of Logistics Engineering at Jacobs University and a Research Affiliate at the Industrial Performance Center (IPC) of the Massachusetts Institute of Technology. Prior to this, he worked as Postdoctoral Research Fellow at the IPC. He holds a doctoral degree in engineering from TU Dortmund University/Fraunhofer IML and another one in logistics from the University of Duisburg-Essen. He studied Logistics Engineering at the University of Duisburg-Essen and Industrial Engineering at Südwestfalen University of Applied Sciences.

Maria Ilie is a research assistant at Jacobs University Bremen majoring in Global Economics and Management. Prior to this, she worked as a Business Analyst for Microsoft UK.

Durch Digitalisierung entsteht Qualität – Produktionsdaten bei Lieferanten für die Qualität des Kundenprodukts nutzbar zu machen

7

Per Larsen

Zusammenfassung

Das vorliegende Kapitel befasst sich mit der Digitalisierung im Umfeld von Qualitätsmanagement. Die hierbei zugrunde liegende Idee ist, dass Produktionsdaten bei Lieferanten digital erfasst werden, um bezogen auf ein spezifisches Kundenprodukt Qualitätsinformationen abzuleiten und für den Kunden nutzbar zu machen. Untersucht wird die Idee durch den Anlagenbauer DISA, der Anlagen für die Herstellung von Eisengussteilen erzeugt. Bei der Herstellung von Eisenguss kann davon ausgegangen werden, dass Qualitätskosten bei Gussabnehmern mit ein Richtwert von zwei bis vier Prozent der Einkaufsumme entstehen können. Zusätzlich entstehen in Höhe von drei bis fünf Prozent Qualitätskosten in der Gießerei selbst. Es ist also ein wesentliches kommerzielles Verbesserungspotenzial vorhanden. Die in Gießereien vorhandenen Systeme zur Optimierung von Produktivität und Qualität können unter Berücksichtigung von Chancen aus der Digitalisierung auch eingesetzt werden, um verbesserte Möglichkeiten zur Reduktion der Qualitätskosten bei Gussabnehmern zu erzielen. Dies kann beispielsweise durch sehr transparente Datennutzung zwischen Gießer und Gussabnehmer realisiert werden, wenn Produktions- und Prozessdaten einer Gießerei mit dem Gussabnehmer geteilt werden. Dies wird durch das Einführen eines zentralen Systems in der Gießerei ermöglicht, das Echtzeitzugang zu Daten aller wichtigen Prozesse der Gießerei erlaubt.

P. Larsen (✉)
Taastrup, Dänemark
E-Mail: per.larsen@noricangroup.com

© Springer Fachmedien Wiesbaden GmbH 2018
F. Schupp und H. Wöhner (Hrsg.), *Digitalisierung im Einkauf,*
https://doi.org/10.1007/978-3-658-16909-1_7

Die Optimierung der Gussqualität durch eine enge Zusammenarbeit zwischen Gießern und Gussabnehmern wird jedoch nicht ändern, dass die Hauptanstrengungen zur Optimierung von Gussqualität weiterhin in erster Linie intern in den Gießereien vorangetrieben werden müssen.

7.1   Einleitung

Das Thema der digitalen Nutzbarmachung von Produktionsdaten zur Verbesserung der Produktqualität bei Kunden wird in diesem Kapitel aus Sicht eines Zulieferers zur Gießereiindustrie beschrieben. Der dänische Zulieferer DISA[1] stellt Anlagen für Eisengießereien her und schafft damit Voraussetzungen für die Erhebung von Produktionsdaten von Eisengusslieferanten. Im Fokus der Betrachtung sind neue Möglichkeiten seitens der Hersteller von in der Gießerei benötigten Maschinen sowie das Verhältnis zwischen Gießereien und deren Kunden.

7.2   Hintergrund

Die Gießereien haben natürlich schon in der Vergangenheit einen laufenden Verbesserungsprozess gehabt, um Produktivität und Qualität zu steigern. Dieser Prozess hat sich in früheren Jahren jedoch weitgehend auf das Wissen der einzelnen Mitarbeiter gestützt. In den letzten Jahren wird der Verbesserungsprozess nun mehr und mehr von Daten unterstützt. Somit findet eine Entwicklung statt von einer Situation, in der die spezifische Gießerei sehr abhängig von den einzelnen Mitarbeiter war, hin zu einer Situation, in der Daten und der Zugang zu Daten wichtiger werden. In vielen Gießereien hat man heute Zugang zu den wichtigen Produktions- und Prozessdaten[2], die die Qualität steuern. Die Herausforderung dabei ist, dass man in fast allen Gießereien keinen Zugang zu den wichtigen Produktions- und Prozessdaten in ein und demselben System hat. Man hat ein System, das die Daten der Sandaufbereitung einsammelt, ein System, das die Daten

---

[1]DISA Industries A/S, für weitere Details siehe: https://www.disagroup.com/en-gb.

[2]Produktions- und Prozessdaten werden hier folgendermaßen definiert: Produktionsdaten sind z. B. eingestellter Pressdruck und Schussdruck einer Formmaschine. Prozessdaten sind z. B. die aktuell gemessene Gießtemperatur, aktuell in der Formmaschine bei Herstellung der Form gemessene Kompaktibilität des Formsands. Alles wird bis zum Formniveau herunter registriert und in einer Datenbank gespeichert.

des Schmelzbetriebs registriert, ein System, das die Daten der Formherstellung aufzeichnet usw. In manchen Fällen sind die registrierten Produktions- und Prozessdaten nicht elektronisch, sondern lediglich in Papierformat vorhanden. Im Endeffekt wird die systematisch laufende Arbeit in der Gießerei, um Produktivität und Qualität zu verbessern, dadurch stark behindert. Es ergibt sich also ein klares Verbesserungspotenzial.

Im Moment wird bei DISA als einem Zulieferer der Gießereiindustrie daran gearbeitet, zentrale Systeme zu entwickeln, die Zugang zu allen wichtigen Produktions- und Prozessdaten in *einem* System schaffen. Das wird typischerweise dadurch gemacht, dass man sich mit den verschiedenen anderen Lieferanten von Ausstattung und Anlagen in Gießereien auf ein Datenprotokoll verständigt. Wichtiger Teil dieses Datenprotokolls ist eine zentrale Uhrzeit, die es möglich macht, sicherzustellen, dass alle Daten korrekt synchronisiert werden. Durch ein solches Datenprotokoll kann man direkt Ursprungsproduktions- und -prozessdaten einsammeln – und zwar von allen Hauptelementen in der Gießerei, wie zum Beispiel Sandaufbereitung, Sandlaboratorium, Schmelzerei, Schmelzbehandlung, metallurgisches Laboratorium, Gießvorrichtung, Impfvorrichtung, Formherstellung, Kernherstellung, Gusskühlung, Gussauspacken, Gussreinigung und Ausschussdaten.

Der direkte Zugang zu allen Produktions- und Prozessdaten in einem System, sowohl in Echtzeit als auch zu historischen Daten, erleichtert die Optimierungsarbeit erheblich. Im Grunde genommen bleibt es zwar immer noch eine manuelle Arbeit, die Ursache, wie beispielsweise für Ausschuss wegen Porositäten, zu finden. Mit all den in einem System vorhandenen Daten eröffnen sich jedoch neue Möglichkeiten. Selbstlernende Systeme werden entwickelt, um die Ursachenzusammenhänge zu identifizieren. Eine Gießerei hat eine sehr komplexe Prozesskette. Daher ist es manchmal für das menschliche Gehirn schwierig, die Zusammenhänge zu identifizieren, die zu Ausschuss führen. Hier war man früher vom Mitarbeiter mit oftmals lebenslanger Erfahrung abhängig. Und dennoch konnten Prozessparameter in Einzelsituationen verbessert, aber nur schwer systematisch abgegrenzt, geprüft und übertragen werden. In den Fällen, wo mehrere Prozessparameter Einfluss aufeinander haben, war es schwer für das menschliche Gehirn mitzuhalten oder in der Praxis oft tatsächlich unmöglich.

Denn die oben beschriebenen Systeme werden in erster Linie entwickelt, um den Gießereien bessere Möglichkeiten zu geben, Produktivität und Qualität zu erhöhen. Dies bedeutet unter anderem, weniger Ausschuss zu produzieren. Optimierung von Qualität wird dabei weiterhin in erster Linie intern in Gießereien vorangetrieben. Jedoch, wenn die Produktions- und Prozessdaten vorhanden sind, eröffnen sich eine Vielzahl neuer Möglichkeiten, auch bei den

Gussabnehmern – also den Kunden der Gießereien – Vorteile aus den vorgela-
gerten Produktionsdaten zu ziehen. Diese Möglichkeiten werden in den folgenden Absätzen beschrieben.

7.3 Möglichkeiten zur Nutzbarmachung von Produktionsdaten

Wie bei den Gussherstellern muss auch bei den Gussabnehmern der Optimie-
rungsprozess weiter vorangetrieben werden, um noch effizienter zu arbeiten. In
der Großserienproduktion ist es bei Eisenguss nicht ungewöhnlich, dass Gussabnehmer Qualitätskosten in Höhe von zwei bis vier Prozent des Einkaufsvolumens
haben. Die Höhe der Qualitätskosten ist stark abhängig von einer Reihe von Faktoren, aber der Mittelwert von drei Prozent könnte nach der Marktsicht von DISA
ein realistischer Richtwert sein. Somit ergibt sich neben dem Qualitätskostenpo-
tenzial beim Hersteller zusätzlich ein wesentliches Einsparpotenzial beim Gussabnehmer.

Wenn man die Qualitätskosten relativ zu Einkaufsummen sieht und mit anderen Ur- oder Umformverfahren wie zum Beispiel Stanzprozessen vergleicht, sind
die Qualitätskosten bei Guss überdurchschnittlich hoch. Um den Gießprozess
noch weiter wettbewerbsfähig zu machen, hat es demnach auch Sinn, verbesserte
Systeme zu entwickeln, um Qualitätskosten zu reduzieren.

Heute ist es generell so, dass der Gießer notwendige Nachweise über Produk-
tions- und Prozessdaten im Falle fehlerhafter Lieferungen bereitstellen können
muss. Mit den neuen digitalen Möglichkeiten, die Zugang zu allen Produktions-
und Prozessdaten elektronisch in einem System ermöglichen, hat der Gießer
nun verschiedene neue Wege, mit denen auch Gussabnehmer Vorteile durch den
Zugang zu diese Daten bekommen können. Zwei dieser neuen Wege werden
nachfolgend beschrieben.

Zunächst muss jedoch konkretisiert werden, von welchem Datenzugang die
Rede ist. Es geht explizit nicht um Produktivitätsdaten, wie zum Beispiel Anzahl
Formen pro Stunde, sondern um Daten, die direkten Einfluss auf die Qualität von
Guss haben.

7.3.1 Gusszertifikat

Die erste Möglichkeit, mit der sich unmittelbar ein effizienter Zugang zu Produk-
tions- und Prozessdaten ergibt, ist die Erzeugung von Gusszertifikaten. Damit ist

gemeint, dass mit jeder Lieferung von Guss ein Zertifikat mitgeliefert wird, das eine Anzahl vereinbarter Produktions- und Prozessdaten auflistet. Die Granularität dieser Daten kann bis auf Formenniveau sein. Zusätzlich können auch vereinbarte Minimal- und Maximalwerte sowie verschiedene statische Daten Teil eines solchen Zertifikats sein.

Mit ein Gusszertifikat können die Gießereien besser gewährleisten, dass die Lieferungen der abgesprochenen Qualität entsprechen, das heißt unter korrekten Parametern produziert wurde. Es entsteht eine größere Sicherheit für weniger Qualitätsprobleme und dadurch mehr Wert in den gelieferten Gussteilen. Die aktuelle Problematik einer Blackbox-Produktion aus Sicht des Kunden wäre somit behoben.

Grundsätzlich bringt dieses Vorgehen produktionsprozessseitig zunächst nichts Neues im Vergleich zu der Ist-Situation, in der die Gießerei für sich selber sicherstellt, dass alle Gussteile unter den Produktions- und Prozessdaten hergestellt sind, die für das Erreichen der abgesprochenen Qualität notwendig sind. Jedoch wird diese neue Möglichkeit sowohl Gießern als auch Gussabnehmern ein Werkzeug in die Hände geben, das hilfreich sein wird, in gemeinsamen Verbesserungsprozessen viel konkreter und zielgerichteter über Probleme, Ursachen und Lösungsansätze bei der Qualitätsverbesserung diskutieren zu können.

7.3.2 Echtzeitlösungen

Zugang zu Daten in einem zentralen System der Gießerei zu erhalten, anstatt eine Vielzahl von Systemen separat betrachten zu müssen, eröffnet auf der einen Seite Möglichkeiten für effektive Echtzeitanalysen in der Gießerei, um Produktivität und Qualität zu steigern, und auf der anderen Seite bietet es gleichermaßen Möglichkeiten für den Echtzeitdatenzugang durch die Gussabnehmer selbst.

Über diesen Weg kann eine sehr transparente Zusammenarbeit von Gießerei und Gussabnehmer entstehen. Daraus ergeben sich mehrere weiterführende Möglichkeiten. Kritische Prozessdaten, wie zum Beispiel die Dauer zwischen einer Behandlung des flüssigen Eisens mit Legierungselementen und dem Abgießen bei der Produktion der Gussteile, können in Echtzeit über eine App in Smartphones für den Gussabnehmer präsentiert werden.

Um gemeinsam, für die Gießerei und den Gussabnehmer, einen Mehrwert durch Echtzeitlösungen zu bekommen, wird ein Echtzeitsystem auch Ansprüche an die Gussabnehmer stellen. Man riskiert dabei, durch die tieferen Einblicke in die Prozessdaten in eine Situation zu kommen, wo der Gussabnehmer enge Toleranzen aufstellt, um auf Nummer sicher zu gehen. Unnötig enge Toleranzen werden dabei aber auch die Kosten in die Höhe treiben. Dieses Preisrisiko wirkt in Kombination mit Echtzeitlösungen gegenseitig regulierend. Sind die Toleranzen zu eng eingestellt, steigt der Preis proportional. Bei entsprechender Rückkorrektur der Toleranzen kann der Preis wieder normalisiert werden.

Die Reise in Richtung Transparenz zwischen Gießer versus Gussabnehmer ist schon gestartet. Ein Beispiel dafür ist die Zusammenarbeit zwischen Gießer und Gussabnehmer, um die Gießbarkeit zu optimieren und damit Kosten zu reduzieren. Voraussetzung für diese Zusammenarbeit ist ein sehr hohes Wissensniveau bei Gießern, das heißt, Gießer sind nicht „nur" Gießer, sondern Mitentwickler der Gussteile. Gießer haben die Anforderungen an die Gussteile im Blick und zeigen Gestaltungsalternativen auf. Um gemeinsame Vorteile aus transparenten Datensystemen zu ziehen, werden analog zu den Anforderungen an den Gießer bei der Optimierung der Gießbarkeit nun Ansprüche an die Gussabnehmer gestellt.

Die sehr transparente Zusammenarbeit macht es nun auch möglich, im Zweifelsfall sofort zu entscheiden, ob Guss aus kritischen Produktionsdatentoleranzlagen ausgeliefert werden soll oder aufgrund von Abweichungen der Prozessparameter keine Freigabe erfolgt. Dabei besteht für die Gussabnehmer die Herausforderung darin, nicht pauschal mit der einfachen Antwort „Nein" zu antworten. Denn das wird langfristig nicht die kosteneffektivste Lösung sein.

7.4 Herausforderungen

Um solche sehr transparenten System zwischen Gießer und Gussabnehmer in Betrieb zu setzen, müssen mehrere Herausforderungen bewältigt werden. Hierzu zählen ökonomische, technische und mentale Herausforderungen.

7.4.1 Ökonomische Herausforderungen bei Gießern

In erste Linie ist eine ökonomische Herausforderung zu bewältigen. Wenn Gießer keinen ökonomischen Vorteil beim Einführen von Systemen erarbeiten können, die zentralen Zugang zu allen für Gussqualität wichtigen Daten geben, wird eine Umsetzung nur schwer realisierbar.

Wie schon erwähnt können zwei bis vier Prozent des Einkaufspreises als Richtwert für Qualitätskosten bei Gussabnehmern angenommen werden. Dazu kommen Qualitätskosten bei Gießereien, die in derselben Größenordnung liegen. Mit Systemen, die Produktivität, Ausschussquote und Ausbringung verbessern, sieht es damit von ökonomischer Seite her vielversprechend aus. Das bedeutet auch, dass Qualitätssysteme den Ausschuss im ersten Schritt nicht vollständig verhindern müssen, um bereits einen signifikanten Verbesserungswert zu erreichen. Klare Aussagen und belegbare Ergebnisse können jedoch im Moment noch nicht vorgelegt werden, da nur sehr wenige Gießereien schon komplett digitalisiert sind.

7.4.2 Technische Herausforderungen

Die meisten technischen Herausforderungen, um digitalisierte Produktionsdaten nutzbar zu machen, sind direkt lösbar. Zum Beispiel werden Cloud-Lösungen sehr schnell entwickelt. Dies bedeutet, dass man Standardlösungen für viele Teilaufgaben verwenden kann, anstatt alles selbst von vorne zu entwickeln. Aber ein Element hat doch eine spezielle Rolle: die Datensicherheit. In erste Linie stellt sich die Frage „Wie gewährleistet man Datensicherheit?" und in zweite Linie „Wie überzeugt man skeptische Gussabnehmer, dass man wirklich Datensicherheit hat?" Die ersten Erfahrungen zeigen, dass das Letztere die größere Aufgabe sein kann. Auf der anderen Seite darf erwartet werden, dass die Digitalisierung von und Hantierung mit Daten den größeren Teil des Unternehmensalltags prägen wird. Die Herausforderungen mit skeptischen Gussabnehmern werden dann weniger werden.

7.4.3 Mentale Herausforderungen bei Gießern

Eine sehr transparente Zusammenarbeit zwischen Gießer und Gussabnehmer wird für manchen Gießer sicherlich grenzüberschreitenden Charakter haben. Ein Aspekt ist das Abgeben von Souveränität. Mit einem sehr transparenten Zugang zu Produktions- und Prozessdaten wird der Gussabnehmer Daten sehen, im Bezug auf Typen und Menge, die bis jetzt noch nie aus einer Gießerei herauskamen. Das wird in der Praxis unvermeidlich den Spielraum der Gießer kleiner machen. Einige Gießer werden zurückhaltend sein. Im Endeffekt steht es natürlich jedem Gießer frei, ob er diese transparente Zusammenarbeit mit Gussabnehmer anbieten will. Hier muss aber auch wieder unterstrichen werden, dass sich die Digitalisierungsbemühungen nicht um Produktivitätsdaten, sondern um Daten drehen, die direkt zur Gussqualität in Beziehung stehen.

7.4.4 Mentale Herausforderungen bei anderen Kunden der Gießerei

Die Einführung einer sehr transparenten Zusammenarbeit zwischen Gießer und einem Gussabnehmer kann ein weiteres Dilemma beim Gießer erzeugen. Wie überzeugt man die anderen Kunden der Gießerei, die nicht diese transparente Zusammenarbeit nachfragen, dass die Daten ihres Gusses sicher sind und nicht durch Fehler oder gesetzwidrigen Zugang dritten Personen zugänglich gemacht werden? Speziell muss man erwarten, dass hier Großserienkunden eine Herausforderung darstellen werden.

Bleibt man bei der Lösung mit einem Gusszertifikat, das mit jeder Lieferung mitgeschickt wird, werden mehrere der oben beschriebenen Herausforderungen klar weniger. Von einem praktischen Blick aus könnte es daher Sinn machen, zunächst ein digitales Gusszertifikat einzuführen.

7.5 Konklusion

Die Qualitätskosten bei Gussabnehmern von Eisenguss sind abhängig von einer Reihe von Faktoren und können mit einem Richtwert von zwei bis vier Prozent der Einkaufsumme angenommen werden. Die Qualitätskosten in Gießereien liegen in einer Größenordnung von drei bis fünf Prozent der Herstellkosten. Es ist also durch Qualitätsverbesserung ein wesentliches Einsparpotenzial vorhanden.

Im Moment werden Systeme entwickelt und implementiert, die Gießereien Echtzeitzugang zu Produktions- und Prozessdaten aller wichtigen Gebiete geben, zum Beispiel in der Schmelzerei, Sandherstellung, Formherstellung, Gießeinrichtung, Impfeinrichtung und beim Auspacken. Diese Systeme werden die internen Prozesse in Gießereien effektiver machen, um Produktivität und Qualität zu verbessern.

Selbstlernende Systeme, die komplexe Zusammenhänge zwischen Produktions- und Prozessdaten und Ausschuss erkennen können, die von Menschen hingegen nur sehr schwer erkennbar sind, sind auch zusätzlich zu implementieren.

Einen transparenten Zugang zu Daten in der Gießer-Gussabnehmer-Beziehung, wo der Gussabnehmern auch Zugang zu vorhandenen Produktions- und Prozessdaten des Lieferanten bekommt, kann den gemeinsamen Verbesserungsprozess stark unterstützen und dadurch auch beim Gussabnehmer zu Reduktionen von Qualitätskosten beitragen.

Der sehr transparente Datenzugang stellt einige Herausforderungen. Diese sind von sowohl ökonomischer also auch technischer und mentaler Natur. Letztendlich wird die gemeinsame Datennutzung den Gießer nicht aus der Pflicht nehmen. Eine sehr transparente Zusammenarbeit zwischen Gießer und Gussabnehmer wird auch nicht ändern, dass die Hauptanstrengungen für das Erreichen und Optimieren der Gussqualität in der Gießerei liegen.

Als Lieferant für die Gießereiindustrie kann man nur helfen, diese Lösungen für die Industrie verfügbar zu machen. Im Endeffekt sind es die Gießereien und deren Kunden, die entschieden, ob der vorgestellte Weg verfolgt wird. Es ist doch für den Verfasser dieses Kapitels eindeutig, dass Zugang zu Produktions- und Prozessdaten wichtiger Bestandteil der Zukunft ist, um in der Gusslieferkette wettbewerbsfähig zu bleiben.

7.6 Ausblick

Der Zugang zu Daten bei Lieferanten, um die Qualität des Kundenprodukts wie beschrieben zu optimieren, kann noch weiterentwickelt werden. Eine Anzahl weiterer Möglichkeiten entfaltet sich. Zum Beispiel können Daten nicht nur von Lieferanten zu Gussabnehmern strömen, sondern auch zurück von Gussabnehmern zu Lieferanten. Daten über Ausschuss, der bei Kunden gefunden wird, wie zum Beispiel Sandeinschlüsse, die erst bei der Bearbeitung entdeckt werden, können zurück zu Lieferanten gemeldet werden. Dadurch werden die Algorithmen, die automatisch zur Prozessoptimierung eingesetzt werden, noch leistungsfähiger.

Über den Autor

Dr. Per Larsen, Product Portfolio & Innovation Manager, geboren 1971 in Odense, Dänemark, hat 1996 als Diplomingenieur sein Studium der Gießereitechnik als Hauptrichtung an der Technischen Universität Dänemark (DTU) abgeschlossen. 1996 wurde er bei DISA Industries A/S als Gießereianwendungstechniker angestellt. Seine Hauptaufgabe war, gießereitechnische Probleme am Ort bei DISAMATIC-Kunden zu lösen. 2006 hat er seine Doktorarbeit „Gießen von dünnwandigen Eisenteilen in vertikal geteilten Formen" fertiggestellt. Über die letzten 10 Jahre hat Dr. Larsen rund 10 Doktoranden betreut. Seit 2006 ist Dr. Larsen Innovation Manager bei DISA Industries A/S, seit 2017 Product Portfolio & Innovation Manager bei DISA. 2016 hat Dr. Larsen das CEL Programm – Certificate in Entrepreneurial Leadership – bei der DTU Business beendet.

Digitalisierung in der Lieferantenanbindung

Heiko Wöhner

Zusammenfassung

Digitalisierung ermöglicht die Verbesserung der Material- und Informationsflüsse zwischen Unternehmen und deren Lieferantenbasis. Dabei werden die Planungs-, Lieferungs- und Produktionsprozesse zwischen Lieferant, Spediteur und Abnehmer durch logistische Lieferantenanbindung verzahnt. Dieses Kapitel beschreibt sieben aufeinander aufbauende Digitalisierungsstufen. In den ersten drei Stufen werden grundlegende Standards für die Digitalisierung geschaffen und bestehende Prozesse durch Digitalisierung effizienter gestaltet. Unter anderem wird die unternehmensübergreifende Sichtbarkeit der tatsächlichen Bestände sukzessive erhöht und ermöglicht verbesserte Steuerungsentscheidungen. Die folgenden zwei Digitalisierungsstufen beschreiben Verzahnungsmechanismen in Bestell- und Lieferverhalten, Transport und Produktion, die erst durch Digitalisierung ermöglicht werden. Beispielsweise wird die Auslastung der Produktions- und Transportkapazitäten zwischen Lieferant, Spediteur und Abnehmer durch dynamische Abstimmung verbessert. Die weiteren zwei Digitalisierungsstufen erhöhen die Effizienz des Planungssystems durch den Wandel in ein System mit dezentraler Datenhaltung und Entscheidungsfindung. Die zeitliche Investition in das Lesen dieses Artikels dürfte für Sie lohnenswert sein, wenn folgende Fragen für Sie interessant und relevant erscheinen: Welche Chancen bieten sich für Unternehmen, durch fortschreitende Digitalisierung im Bereich der Lieferantenanbindung zu profitieren? Gibt es in diesem Zusammenhang typische Stufen, die in einer festgelegten

H. Wöhner (✉)
Baden-Baden, Deutschland
URL: http://www.linkedin.com/in/woehner/

© Springer Fachmedien Wiesbaden GmbH 2018
F. Schupp und H. Wöhner (Hrsg.), *Digitalisierung im Einkauf,*
https://doi.org/10.1007/978-3-658-16909-1_8

Reihenfolge durchlaufen werden müssen? Und anhand welcher Zielgrößen und Beispiele kann Nicht-Einkäufern und Nicht-Logistikern im Unternehmen der Mehrwert einer Digitalisierung in der Lieferantenanbindung erklärt werden?

8.1 Einleitung

Zunächst ein paar Worte zum Begriffsverständnis: Mit logistischer Lieferantenanbindung ist die Verzahnung der Material- und Informationsflüsse zwischen einem Unternehmen und seiner Lieferantenbasis gemeint. Typische Felder der Lieferantenanbindung sind Transport- und Behälterkonzepte, Bestell- und Lieferzyklen sowie Nachrichtenformate und Web-Applikationen für den Datenaustausch. Die Vorteile einer fortgeschrittenen Lieferantenanbindung liegen grundsätzlich einerseits in der verbesserten Reaktionsfähigkeit – Flexibilitätsvorteil – und andererseits in verringerten Kosten durch reibungslose, abgestimmte Prozesse – Kostenvorteil. Dabei sind diese Ziele erfreulicherweise nicht nur auf Kosten des jeweils anderen Ziels zu erreichen, sondern wachsen gemeinsam mit zunehmender Digitalisierung in der Lieferantenanbindung.

Um den Entwicklungsgrad der Lieferantenanbindung in Zusammenhang mit Digitalisierung näher zu betrachten, ist die Verarbeitung eingehender Informationen mit den dadurch unterstützten Materialflüssen eine gute Betrachtungsgröße. Denn ausgehende Informationen, wie zum Beispiel Lieferplanabrufe, und abgehende Materialflüsse, wie beispielsweise Leerbehälter- und Rohteilbereitstellungen, können nur begrenzt durch das sendende Unternehmen verfolgt werden. Die Verantwortung für die Verarbeitung der Informationen und die Steuerung der Materialflüsse liegt bei den Lieferanten. Im Gegensatz dazu laufen von Lieferanten eingehende Nachrichten parallel zum Materialfluss und lassen Verarbeitung und Verwendungsmöglichkeiten zentral durch den Kunden ermitteln. Die Betrachtung der eingehenden Nachrichten ist auch deshalb eine relevante Größe, weil Unternehmen per se auch ohne jegliche Digitalisierung in diesem Bereich auskommen können. Material wird telefonisch oder per Fax bestellt und dann hoffentlich pünktlich und in der richtigen Menge geliefert – Vorausblick auf tatsächliche Lieferungen: praktisch null. Mit etwas Erfahrung und der richtigen Büroausrichtung wissen Sie als Einkäufer oder Logistiker beim Blick aus dem Fenster, welches Material auf dem heranfahrenden LKW sein müsste. Dieser Zustand bildet die Nulllinie für Lieferantenanbindung und deren Digitalisierung. Wenn Ihre Planung eins zu eins von allen Lieferanten umgesetzt wird, könnten

Sie sich mit dieser Stufe zufriedengeben. Falls eine hundertprozentige Umsetzung mangels Wollen oder Können bei einem Teil der Lieferanten nicht möglich ist, oder falls Sie sich Optimierungspotenzial erschließen wollen, das nur durch Digitalisierung erreichbar ist, stellt die Lieferantenanbindung eine attraktive Möglichkeit dar. Denn durch Ausweitung des für Beschaffer sichtbaren Bereichs an sendungsrelevanten Informationen vergrößert sich der Lösungsraum bei Optimierungsaufgaben. Je weiter die Systemgrenzen nach außen verschoben werden, desto besser kann auch das Gesamtoptimum bestimmt und realisiert werden. Hierzu lassen sich sieben Digitalisierungsstufen der Lieferantenanbindung unterscheiden, wobei jeweils der Anbindungs- und Digitalisierungsgrad von Stufe zu Stufe wächst. Die Digitalisierungsstufen können in der Regel nur in aufsteigender Reihenfolge durchlaufen werden. Um also von den neuartigen Digitalisierungschancen der höheren Stufen profitieren zu können, muss zunächst Basisarbeit geleistet werden.

8.2 Digitalisierungsstufe 1: In der digitalisierungsvorbereitenden Stufe werden durch Formalisierung und Standardisierung Voraussetzungen für erfolgreiche Digitalisierung geschaffen

Die Systemgrenze bildet die Beschaffungslieferkette in das eigene Unternehmen: Lieferanten, Spediteure und das eigene Unternehmen als Abnehmer. Ziel dieser Stufe ist nicht die Erweiterung der Systemgrenze, sondern das Verhalten des Systems Lieferant-Spediteur-Abnehmer besser vorherzusagen. Aus Sicht des Einkäufers, der für Cash Flow und Zukaufteilbestände mitverantwortlich ist, lässt sich damit bereits die erste überschüssige Luft aus dem System nehmen. Je verlässlicher das System funktioniert, desto weniger Sicherheitspuffer für Unerwartetes müssen vorgehalten werden. Weniger Sicherheitspuffer verschlanken wiederum das System und steigern positiv selbstverstärkend dessen Steuerbarkeit.

Ein erstes Mittel der Wahl in Digitalisierungsstufe 1 sind Lieferanten-Logistikvereinbarungen. Mit einer Lieferanten-Logistikvereinbarung werden Rechte und Pflichten, gegenseitige Erwartungen und logistische Parameter im Kunden-Lieferanten-Verhältnis geregelt. Typische Inhalte sind die auf Basis der Jahresplanmengen erwartete Flexibilität, die Länge des Vorschauhorizonts und dessen Einteilung in ein oder mehrere Verbindlichkeits- und Flexibilitätszonen samt Abgrenzung eines Fixierungshorizonts, in dem grundsätzlich keine Änderungen mehr vorgenommen werden. Zusätzlich können noch Anliefertage, Transportlaufzeiten,

Behälterkreisläufe etc. vereinbart werden. Eine derartige Abstimmung ist notwendige Voraussetzung für ein akkurates Bestell- und Lieferverhalten.

Stellen Sie sich vor, Sie als Kunde erwarten tagesgenaue Lieferungen entsprechend Ihrer Lieferplaneinteilungen und Ihr Lieferant sieht die per Fax übermittelten Daten eher als grobe Orientierung für den Versand seiner in losgrößenorientierter Produktion fertigwerdenden Materialien. Die Lieferbeziehung abzustimmen und relevante Größen in Form einer Lieferanten-Logistikvereinbarung vor Aufnahme des Lieferverhältnisses festzuhalten, kann die Ursache vieler operativer Lieferprobleme beheben.

Wenn Lieferant und Kunde dasselbe Verständnis von den Erwartungen an die Qualität der Lieferbeziehung haben, ist die Basis für ein harmonisches Bestell- und Lieferverhalten gegeben. In der praktischen Umsetzung wird sich zeigen, ob die Erwartungen auch erfüllt werden. Hierbei ist es erstens positiv, wenn vor Unterzeichnung Bereiche mit zumindest teilweise entgegengesetzt gelagerten Interessen, zum Beispiel Einkauf und Logistik, an der Entstehung beteiligt waren, und zweitens förderlich, die getroffenen Vereinbarungen auch durch entsprechende Systemeinstellungen zu operationalisieren. Dabei hilft es aus Kundensicht, Anforderungen zu standardisieren und mit ähnlichen Lieferanten auch ähnliche Vereinbarungen zu schließen. Eine interne Regelung könnte sein, dass Lieferanten von relativ geringwertigem Kleinmaterial immer nur einmal die Woche anliefern und zwar an einem festen Anliefertag, sagen wir mittwochs. Das Planungssystem übernimmt dann nicht nur vormals manuell ausgeführte Ausgleichstätigkeiten wie Zusammenfassen von Kleinmengen, sondern die Standards vereinfachen auch das Verständnis für die Prozesse. Schraube gleich Kleinmaterial gleich Mittwochsanlieferung. Dieses Beispiel mag für Sie unter der Überschrift „Digitalisierung" merkwürdig klingen, solche Standards erleichtern es aber, die Aufmerksamkeit und Ressourcen auf Wichtigeres zu bringen.

Die Wirtschaftlichkeit steigt in dieser Stufe durch die Verbesserung der Systemeinstellungen und den Entfall nicht wertschöpfender Tätigkeiten. Es gibt etliche Potenziale vor der Digitalisierung, die durch Lieferantenanbindung gehoben werden können, zum Beispiel die Bündelung im Transport durch über mehrere Lieferanten abgestimmte Liefertage. Bestimmt haben Sie eine Landkarte Ihrer Lieferanten an der Wand hängen oder am Bildschirm verfügbar. Gibt es regionale Cluster, in denen mehrere Lieferanten ansässig sind? Werden für diese Lieferanten die Anlieferungen auf dieselben Tage geplant? Und liefern die Lieferanten auch tatsächlich entsprechend? Ein abgestimmtes Lieferverhalten erhöht die Wirtschaftlichkeit der Transporte und rhythmisiert die Wareneingangsprozesse. Über eine Kontrolle der Prozesseinhaltung, beispielsweise in Form einer logistischen

Lieferantenbewertung, und falls notwendig Korrekturmaßnahmen kann das Auftreten von Abweichungen reduziert werden.

8.3 Digitalisierungsstufe 2: Der Einstieg in die Digitalisierung durch Erfassung der nicht mehr änderbaren Lieferungen im Planungssystem des Abnehmers (Lieferschein- und Lieferungsavisierung)

Digitalisierung verschiebt die Systemgrenzen des Planens und Optimierens vom Eintreffen der Sendung auf den Versandzeitpunkt. Bei typischen nationalen Nachtsprungverkehren kann von einer 16 h früheren Information ausgegangen werden. Beschaffer können die Lieferdaten, ob passend zur Bestellung oder von dieser abweichend, für Planungsaktualisierungen am Vortag der Anlieferung nutzen.

Für eine Lieferscheinavisierung sind denkbar wenig Basisdaten notwendig: Lieferscheinnummer, Materialnummer mit Bestellbezug, Liefermenge und erwartetes Eintreffdatum. Je nach Anwendungsfall können noch weitere Informationen, wie beispielsweise die Lieferantencharge, ergänzt werden. Mit einer solchen Lieferscheinavisierung lässt sich zum einen wie beschrieben die Planung vor Eintreffen der Sendung aktualisieren und zum anderen kann die Wareneingangsbuchung vorerfasst werden, sodass bei Eintreffen der Sendung eine beschleunigte Abfertigung möglich ist. Noch schöner ist es, von Lieferanten zu jeder Sendung eine Lieferungsavisierung zu erhalten. In der Lieferungsavisierung wird die Lieferscheinavisierung um behälter- und sendungsstrukturbezogene Daten ergänzt. Es wird beispielsweise mit Versand beim Lieferanten elektronisch übertragen, dass die mit den Nummern 1001 bis 1024 gekennzeichneten Behälter auf einer Palette mit Abdeckplatte angeliefert werden und die ganze Palette die Nummer 1025 trägt. Behälternummern auf den einzelnen Behältern (Single-Label) und der Palette (Master-Label) entsprechen eins zu eins den physisch vom Lieferanten an der Palette angebrachten Etiketten, beispielsweise nach dem internationalen Standard Global Transport Label (GTL). Die Lieferungsavisierung ermöglicht dem Abnehmer die behältergenaue Übernahme in sein Planungssystem, den Entfall einer eigenen Etikettierung und die automatische Leergutbuchung zugunsten des Lieferanten. Die Daten im Lieferungsprozess werden durchgängig von der Produktion beim Lieferanten bis zum Verbrauch beim Kunden verwendet und fördern beispielsweise die Transparenz bei Auffälligkeiten der Teilequalität.

Mit der Lieferscheinavisierung werden die Planungsprozesse wirtschaftlicher, weil ein Teil der Nachfragen von Disponenten schlicht entfällt. Wurde das Material gesendet? Kommt auch die gesamte eingeteilte Menge? Wann ist die Ankunft vorgesehen? Disponenten sind im System über die In-Transit-Mengen informiert und können auch die Produktionsplanung auf dieser Basis verbessern. Mit der Lieferungsavisierung steigt die Effizienz im Wareneingang und der Materialfluss wird durchgängig rückverfolgbar. War es vormals notwendig, alle Behälter erst zu einem Erfassungsplatz zu fahren und zu etikettieren, können die angelieferten Materialien nach Scannen des Labels direkt an den Lagerort oder Verbrauchsort gebracht werden. Beim Kunden entfallen Handlingschritte, in der Supply Chain entfällt Doppelarbeit.

8.4 Digitalisierungsstufe 3: Verschieben der Sichtbarkeit von der Abholung auf den Transportanmeldezeitpunkt

Erfolgte die Kommunikation bisher zwischen Lieferanten und Spediteuren jeweils bilateral, ist durch Digitalisierung der Transportanmeldung in Digitalisierungsstufe 3 auch der Abnehmer über ausstehende Transporte im Bilde. Im typischen nationalen Landverkehr ergeben sich so weitere 24 h mehr Transparenz für den Abnehmer. Gleichzeitig schafft dieser Schritt eine wichtige Voraussetzung für die Folgestufen: den Anschluss der Spediteure beziehungsweise Logistikdienstleister an das Planungssystem des Abnehmers. Wie bei Lieferanten ist auch hierbei die Basisanbindung komplett, wenn ein jeweils bidirektionaler Informationsaustausch Kommunikation zwischen Lieferanten, Spediteuren und Abnehmer erlaubt. Die Digitalisierungsstufe 3 ermöglicht dem Beschaffer zudem erstmals, durch die bessere Sicht auf Stellgrößen außerhalb des eigenen Unternehmens Einfluss zu nehmen. Mit der Information über beauftragte, aber noch nicht an Dienstleister vergebene Transporte können Beschaffer die Wahl der Transportart beeinflussen. Um dies auch in der Realität praktikabel zu halten, ist eine feste und begrenzte Auswahl an vordefinierten Alternativen notwendig. Die Kunst dabei ist, ein Optimierungsniveau zu wählen, dass für die Entscheider noch verständlich ist. Standards und Rhythmus des Lieferprozesses dürfen nicht in der Beliebigkeit des theoretisch Möglichen aufgehen.

Man kann unterschiedlicher Meinung sein, ob die dynamische Planung des Transportnetzwerks im Kurzfristhorizont zu den Aufgaben eines Kundenplanungssystems gehört. Aber die Grenzen dieser Optimierung durch einzelne Versender beziehungsweise Empfänger sind durch die Grenzen des Optimierungsbereichs dieser Unternehmen sowie deren Fachexpertise gegeben: Jeder Anwender kann nur auf seiner Datenbasis optimieren. Dabei haben Spediteure lieferkettenübergreifende Optimierungsmöglichkeiten. Selbst bei Komplettladungen stellt sich immer die Herausforderung der Rückladung. Zusätzliche Restriktionen und Variabilisierungsmöglichkeiten wie Lenkzeit, Teilzeitarbeitskräfte, Fahrzeuggrößen, Umlaufgestaltung etc. zu berücksichtigen, erfordert Expertise und ist nicht Kerngeschäft der Versender und Empfänger. Schließlich erhöht sich schlicht die Komplexität durch das Betrachten zusätzlicher Optimierungsvariablen mit abnehmendem Grenznutzen.

Denn auch in einem statischen Transportnetzwerk lassen sich wesentliche Potenziale durch Digitalisierung und Optimierung heben. Dies ist Hausaufgabe der den Transport beauftragenden Unternehmen – und es lohnt sich. Auf Basis einer strategischen Routenoptimierung werden entweder fixe Zuordnungen je Relation vergeben oder mit einer begrenzten Anzahl an Alternativen wie Komplettladung, Teilladung oder Paketdienst kombiniert. Sendungen von Lieferanten an Kundenstandorte werden durch diese Vorgaben zum Beispiel täglich konsolidiert und an Spediteure vergeben. Die erste Bündelung erfolgt durch die Verlader, die zusätzliche Auslastungsoptimierung obliegt in diesem Szenario den Spediteuren.

Die Spediteure erhalten Transportaufträge in standardisierter Form zu einem definierten Zeitpunkt für alle Transporte im Lieferanten-Kunden-Netzwerk. Nach Abholung, bei Zustellung und je nach Transportdauer bei weiteren relevanten Ereignissen setzen die Spediteure Statusmeldungen an das Transportsystem. So ist der Empfänger immer über den Fortschritt der Sendungen informiert. Und wer sich auf das Eintreffen der Sendungen verlassen kann, lässt den Sicherheitsaufschlag in der Planung gerne entfallen. In-Transit-Bestände können so durch eine enge und verlässliche Planung reduziert werden. Mit Abgleich zwischen Soll und Ist können Abweichungen automatisch erkannt und den Entscheidern Alternativen vorgeschlagen werden. Wenn Ihre Sendung im Gebietsspeditionsnetzwerk die Abfahrt des Hauptlauf-LKW verpasst hat, können Sie entscheiden, ob auf die nächste Abfahrt gewartet werden kann oder ein Eiltransport notwendig ist. Sendungen lassen sich im Zusammenspiel aus Kunde und Spediteur auch während des Transports noch beschleunigen. Durch solche Rückkopplungspunkte steigt die Effizienz im Transportprozess.

Die hauptsächlichen ökonomischen Vorteile liegen im mittelfristigen Optimierungsbereich. Wenn alle Transporte von Lieferanten zu Empfangswerken erfasst sind, berechnen gängige, am Markt heute längst verfügbare Transportoptimierungsprogramme eine optimales Set-up für das strategische Transportnetzwerk. Umschlagspunkte und Speditionsgebiete können sinnvoll gewählt werden. Raten

mit Speditionen beruhen dann auf den tatsächlich zu transportierenden Mengen
und können entsprechend genauer verhandelt werden, Sicherheitsaufschläge entfallen. Spitzen im Transportbedarf können erkannt, hinterfragt und zumindest
teilweise behoben werden.

8.5 Digitalisierungsstufe 4: Die Digitalisierung ermöglicht eine dynamische Abstimmung in den Versand- und Transportprozessen

Im Versand kann durch Meldung der versandbereiten Bestände eine automatisierte Kommunikation mit der Kundenplanung erfolgen, ob noch nicht eingeteil-
tes, aber bereits fertiggestelltes Material bereits gesendet werden soll. Innerhalb
des Fixierungshorizonts entsteht so Flexibilität, ohne in die Produktionsprozesse
oder die Transportarten einzugreifen. Unter der Annahme, dass Lieferanten zur
Gewährleistung termingerechter Lieferungen eine Bedarfsvorlaufzeit bei der
Terminierung von Fertigware einplanen, steht dieser Puffer von zum Beispiel
einem Tag als zusätzliche Flexibilität zur Verfügung. Hinsichtlich der Transport-
prozesse erfolgte die Prüfung auf Routenanpassung in Digitalisierungsstufe 3
(Abschn. 8.4) lediglich sporadisch durch Eingreifen des Beschaffers. Im Fall von
Maschine-zu-Maschine-Kommunikation zwischen der Transportanmeldung des
Lieferanten und der Produktionsplanung des Kunden ist für jeden Transport eine
Optimierungsprüfung gegenüber der Echtzeitplanung des Kunden möglich. Die
Produktions- und Transportplanungssysteme gleichen Informationen automatisiert ab und passen die Transportprozesse bei Bedarf entsprechend der aktuellen
Kundenbedarfe an. Hat sich der Bedarf zwischenzeitlich nach vorne verschoben
und fordert eine beschleunigte Lieferung? Oder wurde der Bedarf nach hinten
verschoben und erlaubt es, dem Spediteur die Differenztage als zusätzliche Spiel-
masse zur Fahrzeugauslastung anzubieten?
    Entsprechend liegen die ökonomischen Vorteile dieser Digitalisierungsstufe
in der Produktion des Abnehmers oder bei der Speditionseffizienz. Material, das
schon beim Lieferanten verfügbar, aber eigentlich noch nicht für die Produktion
vorgesehen ist, kann systematisch genutzt werden, um die eigene Produktion
dichter am Kundenbedarf zu fahren. Zusatzbedarfe in der eigenen Produktion
können ohne Nachfragen bei Lieferanten auf Machbarkeit geprüft, eingeplant und
mit den Regelprozessen realisiert werden. Das Material kann schneller fließen.
Lieferanten profitieren, weil ihre produktionslosgrößenbedingten Fertigwaren-
bestände systematisch für Kunden verfügbar sind und teilweise früher verkauft
werden können. Ein Flexibilisierungselement erlaubt es Spediteuren in neuartiger

Weise, Transporte auszulasten. Nicht jeder Lieferant liegt nahe einem der logistischen Hotspots. Bei Lieferanten in ländlicheren Gebieten oder am Rande eines Speditionsgebiets ist es für Spediteure interessant, Material auch einen Tag früher abholen und dann zwischenpuffern zu können, wenn eine Tour in die Region ohnehin geplant und Kapazität vorhanden ist. Die Auslastung der Fahrzeuge steigt und das Anfahren ungünstiger Standorte kann in ein größeres Portfolio an Touren bestmöglich eingeplant werden. Auslastungssteigerung ist damit auch gleich CO_2-Einsparung.

8.6 Digitalisierungsstufe 5: Die Digitalisierung der Lieferantenanbindung ermöglicht neuartige Prozesse in der Produktionsoptimierung – übergreifend zwischen Abnehmer und Lieferanten

Fertigung im eigenen Unternehmen und Fertigung bei Lieferanten werden dynamisch, aber regelbasiert miteinander verzahnt. Dies kann sowohl hinsichtlich der Produktionsplanung als auch hinsichtlich der einzelnen, fertigungsbezogenen Produkteigenschaften erfolgen. Eine weitere neuartige Funktion lässt sich durch die Einbindung produktionsbezogener Daten in die logistische, unternehmensübergreifende Kommunikation schaffen. Typischerweise unterliegen Fertigungsprozesse gewissen Schwankungen, zum Beispiel in der chemischen Zusammensetzung oder der physikalischen Abmessung der Produkte. Toleranzgrenzen für Fertigungsparameter spezifizieren die Produkteigenschaften, lassen aber auch unterschiedliche Lagen innerhalb der Toleranzbreite zu. Wenn Lieferanten auf Behälterebene Produkteigenschaften an Kunden übermitteln, kann der Kunde die nächsten Fertigungsschritte noch feiner auf die Produkte abstimmen. Es ist dann beispielsweise möglich, in der Systemfertigung jeweils solche Komponenten auszuwählen, die ideal zueinander passen. Die Präzision der Systeme steigt.

Hinsichtlich logistischer Fertigungsbeeinflussung kann die Planung auf Detailebene in Echtzeit mit den Bedarfen des Kunden abgestimmt werden. Hierbei können beispielsweise Fertigungslosgrößen und Fertigungsreihenfolge beeinflusst werden. Der Lieferant plant seine Fertigung auf Basis der Mittelfristabrufe, reserviert Kapazitäten für Kunden und beschafft das Vormaterial entsprechend. Hierfür dienen auch die Absicherungen durch die Lieferanten-Logistikvereinbarung. Bevor ein konkretes Produkt für die Fertigung freigegeben wird, prüft das Planungssystem des Lieferanten mit dem Planungssystem des Kunden, ob Stückzahl oder Priorisierung geändert werden sollten. Anhand vordefinierter Optionen werden Grenzen für die Änderungen gesetzt. Wenn der Kunde beispielsweise 1000 Stück von Produkt A bestellt hat, eine wirtschaftliche Fertigung ab 800 Stück möglich ist und das Wartungsintervall des Werkzeugs maximal 1300 Stück beträgt, fragt das Planungssystem des Lieferanten kurz vor Einlastung des Auftrags und unter Berücksichtigung der Vormaterialverfügbarkeit beim Planungssystem des Kunden an, ob die 1000 Stück weiterhin die ideale Liefermenge sind. Durch den Bezug zur Lieferanten-Logistikvereinbarung ist eindeutig geregelt, dass eine Lieferverpflichtung nur für die 1000 Stück gilt. Im Abstimmungsprozess ohne Rückkopplung würden die Planungssysteme aber davon ausgehen, dass diese 1000 Stück auch nicht verändert werden können. Dies entspricht aber häufig nicht der Realität. Bedarfsveränderungen können durchaus im Interesse beider Parteien sein: Einerseits ist der Lieferant gegebenenfalls bei der Rohmateriallieferung an Mindestbestellgrößen oder physikalisch sprungfixe Größen gebunden und hat einen Vorteil, das Material schnell durch die Fertigung zu bringen. Andererseits erlaubt eine Kürzung der Fertigungsmenge innerhalb des für Lieferanten wirtschaftlichen Bereichs, mehr unterschiedliche Produkte im selben Zeitraum auf einer Maschine fertigen zu können und so die Produktionsflexibilität zu erhöhen.

Für den Kunden sind die Änderungen in dem beschriebenen Modus ohnehin vorteilhaft, da er nach Änderungswünschen angefragt wird und Änderungen folglich nur eingeplant werden, wenn sie für den Kunden auch sinnvoll sind.

Fertigungsplanung und -ausführung sowie Logistikplanung und -ausführung sind so miteinander verzahnt. Diese Digitalisierung in der Lieferantenanbindung ermöglicht eine verbesserte Anpassung der Produktionsmengen unterschiedlicher Wertschöpfungsstufen an den aktuellen Kundenbedarf. Mit der Einbindung von Rohmaterial-, Halbfertig- und Fertigmaterialbeständen sowie der tatsächlichen Kapazität beziehungsweise Anlageneffektivität auf Lieferantenmaschinen kann die Kapazitätsplanung dynamisiert werden. Nehmen wir an, im beschriebenen Beispiel hätte der Lieferant zum Zeitpunkt des Abgleichs mit Ausbringungsproblemen zu kämpfen und droht, mit der Bearbeitung seines Fertigungsauftragsportfolios in Verzug zu geraten. Die dem Kunden in genau diesem Fall übermittelte Mindestlosgröße könnte von 800 Stück auf zum Beispiel 400 Stück gesenkt werden. Falls diese Menge für den Kunden ausreichend wäre, würde Kapazität für andere Aufträge geschaffen, sodass auch eine eigentlich zu geringe Fertigungslosgröße sinnvoll sein kann.

Die dynamische Abstimmung der Fertigung bietet somit für Kunde und Lieferant wirtschaftliche Vorteile. Überbestände werden durch die Möglichkeit zur abgestimmten Reduzierung vermieden, Produktionskapazitäten können bei akuten Bedarfen auch durch Steigerungen besser ausgelastet werden. Dabei wird der jeweils kurz vor Fertigungsstart vorliegenden Situation Rechnung getragen. Maschinen und Systeme mit schwankender Anlageneffektivität können kundengerecht beplant und betrieben werden. Flexibilität und Kosten werden gleichzeitig verbessert.

8.7 Digitalisierungsstufe 6: Transfer der Datenhaltung von der zentralen Planungseinheit auf die physischen Behälter

Dies ist der Einstieg in die dezentrale Steuerung der Supply Chain. Behälter sind über eindeutige Nummern voneinander unterscheidbar, beispielsweise durch die Licence Plate des Global Transport Labels. Bei elektronischer Behälterkennzeichnung mit Transpondern können je nach Anwendungsfall und verwendeter Technologie zusätzlich weitere Informationen wie Temperaturverläufe, Beschleunigungswerte und Umgebungsfeuchtigkeit aufgezeichnet werden. Diese Funktionen werden in vielen Pilotanwendungen unterschiedlicher Lieferketten schon heute erprobt. In Digitalisierungsstufe 6 sind derartige Informationen nicht nur verfügbar und werden zu Kontrollzwecken an definierten Punkten ausgelesen, sondern sie sind auf Basis der vorherigen Stufen auch sofort planungsrelevant für die Abstimmung der Transport- und Produktionsplanung sowie für das Qualitätsmanagement.

Nehmen wir an, Sie beziehen Material von einem Lieferanten, dessen Fertigungsprozess mit Verschleiß des Werkzeugs einhergeht. Fertigungslosgrößen sind auf Standzeit und Wartungsintervall des Werkzeugs abgestimmt. Wenn das Befüllen der Behälter beim Lieferanten mit Zeitstempeln protokolliert wird und Parameter des Transports aufgenommen werden, kann die Qualitätsprüfungslogik im Wareneingang verfeinert werden. Typischerweise erfolgt zum Beispiel im Automotive-Umfeld nicht für jede Anlieferung eine Wareneingangsprüfung. Über ein dynamisches Verfahren zur Stichprobenprüfung – Skip-Lot-Verfahren – werden Stichprobenintervalle durch die Ergebnisse der vorherigen Prüfungen gesteuert und bei fehlerfreien Anlieferungen erweitert, um den Gesamtprüfaufwand zu reduzieren. In der durch Skip-Lot sporadisch erzeugten Prüfung könnten gezielt Behälter vom Anfang, der Mitte und dem Ende des Lieferantenproduktionsloses ausgewählt werden. So kann der Kunde nicht nur die Einhaltung der Zeichnungsmerkmale prüfen, sondern auch Informationen über den Fertigungsverlauf des Lieferanten gegen die Erwartungen spiegeln. Es ist zusätzlich möglich, transportbezogene Störungen als Auslöser für eine Qualitätsprüfung zu verwenden. Wurde

während des Transports stark gebremst und könnten Teile beschädigt sein? Oder
gab es einen Temperatur-/Feuchtigkeitsverlauf, der korrosionsfördernd ist? Ein
intelligentes Planungssystem wird diese Qualitätsprüfungen nicht nur bei Pas-
sieren des Wareneingangsgates auslösen, sondern schon vorher durch Abruf der
am Behälter verfügbaren Informationen einplanen. Wenn das Planungssystem
so gestaltet ist, dass im datenbasiert ausgelösten Bedarfsfall jederzeit Warenein-
gangsprüfungen stattfinden können, muss für diesen Prozess auch ein zeitlicher
Puffer zwischen Anlieferzeitpunkt und Verwendung in der Produktion eingeplant
werden. Solch ein Puffer lässt sich stark reduzieren oder gar komplett vermeiden,
wenn über die Behälterinformationen schon während der Transportvorbereitungs-
und Durchführungsphase eine Vorplanung erfolgt. Bei tagesgenauen Lieferungen
könnte mit dem Spediteur vereinbart werden, dass nach auffälligen Situationen
der Transport auf eine 10-Uhr-Lieferung umgestellt werden muss, sodass bis zur
eigentlich spätesten Anlieferung um 17 Uhr auch die zusätzlichen Prüfungspro-
zesse durchgeführt werden können. Der Spediteur kann mit hoher Qualität seiner
Dienstleistung in puncto Beschleunigungs- und Temperaturverlauf Einfluss auf
die Anzahl früher zuzustellender Sendungen nehmen. Gute Qualität beruhigt die
Supply Chain, ein positiv selbstverstärkender Effekt.

Auch für die Skip-Lot-Prüflogik ergibt sich mit solch einer Prüfplanung ein
Vorteil. Eine Auswahl von Prüflosen, die rein auf statistischen Merkmalen beruht,
könnte durch die Dringlichkeit der Verwendung beeinflusst werden. Es wird nicht
eine konkrete Anlieferung zur Prüfung ausgewählt, sondern eine Serie von zum
Beispiel fünf Anlieferungen. Für diese Anlieferungen wird jeweils bei Buchung
des Wareneingangs die Kapazität der Wareneingangskontrolle gegen die Produktionsplanung geprüft. Mit einer Planung der Wareneingangskontrolle als Ressource können Prüfungen so terminiert werden, dass kein zusätzlicher Puffer für
eventuelle Prüfungen vorgehalten werden muss.

Nach Transport und Fertigung kann so mit der dauerhaften Übertragung von
am Behälter gesammelten Daten ein weiterer Prozess optimiert werden: die
Qualitätskontrolle. Nun fragen Sie sich, ob diese zum Beispiel im Automotive-
Umfeld nicht schon vor Jahrzehnten abgeschafft wurde? Teilweise ja, meistens
bei einer Null-Fehler-Orientierung zu Recht, aber mit Blick auf die Rückrufquo-
ten von 51,6 Mio. Fahrzeugen in 2016 in den USA[1] vielleicht zu pauschal. Es
wäre ein veraltetes System, jede Sendung zu prüfen oder hin und wieder eine
Stichprobe zu ziehen. Aber wenn die Auswahl der Prüfungen dem Ausschluss

---

[1]Berechnungen des Center of Automotive Management in Bergisch Gladbach.

konkreter, faktenbasierter Risiken dient, ergibt sich inhaltlich für die Qualitätskontrolle eine neue Qualität. Gleichzeitig können durch Vorausplanung der Ressource Qualitätskontrolle samt der untergeordneten Ressourcen wie 3-D-Messtechnik, Tomografie, Labor etc. Durchlaufzeiten, Auslastung und somit Effizienz gesteigert werden. Es hat durchaus Sinn, Prüfkriterien fallbezogen anzupassen. Wenn die Transportmessung eine korrosionsfördernde Umgebung meldet, muss ein ansonsten unauffälliges und nicht zur Prüfung vorgesehenes Produkt nicht zusätzlich auf zum Beispiel die richtige Legierung geprüft werden. Die Produktionsplanung beim Kunden wird stabilisiert, weil Störungen durch die Reduzierung und Verkürzung von Prüfprozessen verringert werden.

8.8 Digitalisierungsstufe 7: Die am Behälter verfügbaren Informationen erlauben eine Selbststeuerung des Transportsystems

Die Planungslogik wird von zentral auf dezentral umgekehrt. Beim Begriff „dezentrale Datenhaltung" in der Digitalisierungsstufe 6 (Abschn. 8.7) haben Sie es vermutlich schon geahnt: Die Selbststeuerung als Paradigmenwechsel der Planung stellt die Digitalisierungsstufe 7 dar. Mit den Abhängigkeiten von Kundenbedarfen, Transportkapazitäten, Fertigungsauslastung, Prüfnotwendigkeit und deren ständig variierenden Entscheidungsparametern wird die Lieferkette nicht nur potenziell flexibler und effizienter, sondern für den menschlichen Planer im Kontrollturm trotz aller Regeln auch zunehmend unbeherrschbar. Zentrale Großrechner können das Optimierungsproblem übernehmen, stoßen aber in der Effizienz der Abarbeitung an ihre Grenzen. Die schon seit einigen Jahren in diversen Anwendungsgebieten entwickelten selbststeuernden Systeme stellen eine tragfähige Weiterentwicklung dar. Maschinenbelegung, Transportnutzung und Prüfprozesse werden zwischen den Aufträgen und Behältern vor Ort ausgehandelt und im Systemausschnitt entschieden.

Es wird somit auch einfacher, an unterschiedlichen Stellen innerhalb einer Supply Chain verschiedene, jeweils auf die Entscheidungssituation zugeschnittene Kriterien heranzuziehen. Die bis zur Digitalisicrungsstufe 6 entwickelten Messgrößen können in Ausschnitten des Netzwerks um weitere Größen ergänzt werden. Ein Beispiel ist das Behältergewicht und die Stapelbarkeitsinformation bei der Transportkonsolidierung. In Produktion und Planung aufgrund teilespezifischer Verpackungsvorschriften als fix definiert, spielen diese Größen bei der LKW-Beladung eine Rolle. Entsprechend können sie – und zwar in Kombination mit anderen Faktoren wie Dringlichkeit und Empfindlichkeit – das Transportoptimierungsproblem vor Ort lösen. Für die sich anschließenden Prozesse in der werksinternen Logistik sind die hier betrachteten Faktoren dann nicht mehr entscheidungsrelevant. In Kombination mit der Datenerfassung während des Transports können sich die Mischbarkeit von Gütern auf einem LKW und die Entladungsreihenfolge ändern. Stellen Sie sich zum Beispiel das klassische Stauende hinter der Kurve vor. Der Hauptlauf-LKW muss abrupt stark bremsen, der Schocksensor im LKW nimmt den Beschleunigungswert auf und verknüpft ihn mit den Behältern im LKW. Für jeden Behälter wird berechnet, ob eine Prozessstörung vorliegt oder der Wert innerhalb der für das Material tolerablen Grenzen liegt. Sodann werden beim Umschlag auf den Nahverkehrs-LKW erstens Behälter mit schockanfälligem Material von Behältern mit unauffälligen Messdaten getrennt. Es könnte sogar eine Sperrfläche im LKW gekennzeichnet werden. Zweitens werden die Behälter mit schockanfälligem Material so auf den Nahverkehrs-LKW geladen, dass sie direkt bei der Qualitätsprüfung entladen werden. Eine Vermischung wird frühestmöglich vermieden. In einem anderen Beispiel erhalten Behälter unter leicht korrosionsfördernden Bedingungen ein kurzes Mindesthaltbarkeitsdatum. Die Behälter übersteuern nach Ankunft im Wareneingang mit oder ohne Wareneingangsprüfung das First-In-First-Out-Prinzip und werden schnellstmöglich verwendet. So kann nach der Produktion beim Kunden rasch eine neue Konservierung durchgeführt werden. Ein Sonderprozess „zusätzlicher Korrosionsschutz vor Montage" wird gespart und durch den Standardprozess abgedeckt.

Auf diese Weise werden Prozesse und Planung innerhalb der Lieferkette effizienter – und damit auch das Liefernetzwerk an sich. Entscheidungen werden dezentral anhand der für die Folgeprozesse relevanten Größen getroffen und können auch situativ angepasst werden.

8.9   Ein zusammenfassender Blick nach vorn

Die Herausforderung für die Digitalisierung besteht darin, die Balance zwischen der Utopie einer zentral gesteuerten menschenleeren Fabrik der 1990er Jahre und dem bis ins Detail frei optimierenden, dezentral selbststeuernden System zu finden. Denn die Planungs- und Ausführungssysteme werden in absehbarer Zeit immer nur einen Teil der entscheidungsrelevanten Parameter abbilden können. Daher kann die Digitalisierung einen Teil der Optimierungsaufgaben, die heute von Beschaffern übernommen werden, automatisiert und flächendeckend abbilden und einen weiteren Teil neuartiger Optimierungen realisieren. Für die Akzeptanz und damit die Einführungsgeschwindigkeit und den tatsächlichen

Durchdringungsgrad von Digitalisierung in der Lieferantenanbindung ist es aber essenziell, dass Einkäufer und Beschaffer die zugrunde gelegten Mechanismen verstehen und an kritischen Punkten eingreifen können. Komplexitätsreduzierung ist und bleibt wichtig.

Die Digitalisierungsstufen 1 bis 7 beschreiben einen Weg, um ausgehend vom Digitalisierungsgrad null in sinnvoller Reihenfolge Maßnahmen zu ergreifen, die die Supply Chain in ein integriertes, selbststeuerndes System überführen (Abb. 8.1). In der digitalisierungsvorbereitenden Stufe werden durch Formalisierung und Standardisierung Voraussetzungen für erfolgreiche Digitalisierung geschaffen. In den Digitalisierungsstufen 2 und 3 wird die Sichtbarkeit der tatsächlichen Lieferungen durch Lieferschein- und Lieferungsavisierung sowie den Anschluss der Logistikdienstleister an das Kundenplanungssystem erhöht. Die Digitalisierungsstufen 4 und 5 ermöglichen neuartige Funktionen in den Versand-, Transport- und Produktionsprozessen. Auf dieses hohe Anbindungsniveau setzen die Digitalisierungsstufen 6 und 7 auf und heben die Effizienz der Planung durch dezentrale Datenhaltung und die Qualität der Entscheidungsfindung durch selbststeuernde Systeme auf ein neues Level. Und im Gegensatz zu diesem Artikel ist die Digitalisierung in der Lieferantenanbindung damit sicherlich noch nicht an ihrem Ende.

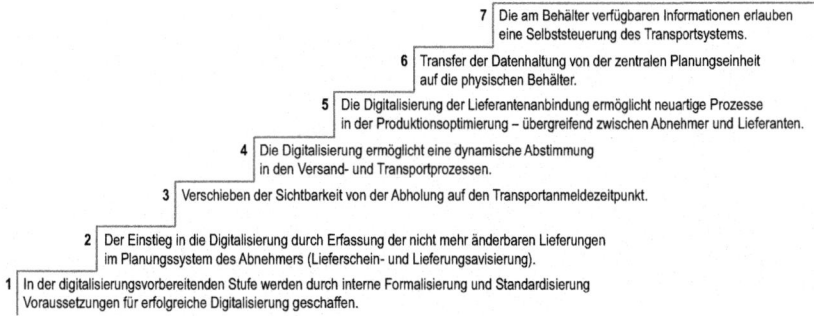

7 | Die am Behälter verfügbaren Informationen erlauben eine Selbststeuerung des Transportsystems.

6 | Transfer der Datenhaltung von der zentralen Planungseinheit auf die physischen Behälter.

5 | Die Digitalisierung der Lieferantenanbindung ermöglicht neuartige Prozesse in der Produktionsoptimierung – übergreifend zwischen Abnehmer und Lieferanten.

4 | Die Digitalisierung ermöglicht eine dynamische Abstimmung in den Versand- und Transportprozessen.

3 | Verschieben der Sichtbarkeit von der Abholung auf den Transportanmeldezeitpunkt.

2 | Der Einstieg in die Digitalisierung durch Erfassung der nicht mehr änderbaren Lieferungen im Planungssystem des Abnehmers (Lieferschein- und Lieferungsavisierung).

1 | In der digitalisierungsvorbereitenden Stufe werden durch interne Formalisierung und Standardisierung Voraussetzungen für erfolgreiche Digitalisierung geschaffen.

Abb. 8.1 Digitalisierungsstufen in der Lieferantenanbindung

Über den Autor

Dr. Heiko Wöhner ist Spezialist Supply Management beim Automobilzulieferer LuK GmbH & Co. KG. Nach seinem Wirtschaftsingenieurstudium in Bremen und Östersund untersuchte er im Rahmen seiner Promotion an der EBS Universität für Wirtschaft und Recht in Wiesbaden, inwiefern Integration mit Kunden und Lieferanten für Unternehmen vorteilhaft ist. Dr. Wöhner sammelte vier Jahre Erfahrungen im Projektmanagement der Bundesvereinigung Logistik (BVL) e. V. und war unter anderem für die inhaltliche Gestaltung des Deutschen Logistik-Kongresses und des Branchenforums Automobil-Logistik mitverantwortlich. Seit 2011 gestaltet Dr. Wöhner die Lieferantenanbindung bei der LuK GmbH & Co. KG und unterstützt Forschung im Bereich Supply Management.

Supplier innovation can be measured – How digitalization allows to effectively include the technology dimension into sourcing decisions

9

Florian Schupp und Matthias Rehm

Abstract

Purchasing supports the development of the technical capabilities of suppliers. In addition purchasing can find and internalize complete new products and product attributes presented by suppliers. Herein the digitalization allows a practical and at the same time fast measurement and valuation of technical and functional value added by innovative product features and technical characteristics. With this, the technology dimension through capturing a product characteristics utility value can be integrated to the supplier sourcing decision in addition to the price, which of course is used in the sourcing decision in form of total cost of ownership. In this chapter it is explained why product innovation by suppliers is element of the scope of purchasing. After addressing several pitfalls in innovation management with suppliers, a model is developed how utility values of product characteristics combinations can be transferred into a preference curve of the buying firm taking reservation prices

F. Schupp (✉)
Bühl, Deutschland
E-Mail: schupp-florian@t-online.de

M. Rehm
Tuttlingen, Deutschland
E-Mail: rehmmatthias@web.de

© Springer Fachmedien Wiesbaden GmbH 2018
F. Schupp und H. Wöhner (Hrsg.), *Digitalisierung im Einkauf,*
https://doi.org/10.1007/978-3-658-16909-1_9

into consideration. As a next step the supplier offers are translated into product characteristics utility-price combinations that are presented in a feasibility curve. Preference curve and feasibility curve are compared with the customer target price. This creates a solution space for supplier innovations below and above the target price.

9.1   How to effectively include the technology dimenison into sourcing decisions

Innovation is driven by ideas. Ideas are generated by individuals or companies. If individuals or companies want to sell their innovations, they need to find a market that wants to buy the innovation. This requires that the product attributes of the innovation are explained to potential customers. In situations where the innovating body exactly meets the technical customer demand, the price will decide if the transaction will happen. In cases where the technical customer demand does not exactly match with the offered product, the specific characteristics of the different product attributes have to be changed to match with possible or acceptable solutions for the customer. When a product has more than one defining attribute, each difference in attribute characteristics results in differentiable product stimuli.

Such stimuli can be ranked by the customer in order to indicate preferences of one product solution over another one. By the method of conjoint analysis following the weighting of product attributes in importance for the customer, the technical utility value of each attribute characteristic can be calculated. The customer can now indicate his payment willingness in form of a reservation price for each product stimulus. The resulting curve out of this activity is a customer preference curve that contains the two dimensions price and technical utility value.

Now, the innovating body has the possibility to offer different product solutions by adjusting the technical characteristics within commonly defined product attributes. Each offered product stimulus can be priced differently, resulting in a feasibility curve by the innovating body. Consequently, all utility-price combinations that are located on or above the customer preference curve are element of all acceptable solutions.

The innovating body can now negotiate not only over price, but also over technical utility. In effect, innovation gets a measurable value that can be used in negotiations or business decisions such as sourcings. At the same time, wrong product-price allocations are eliminated.

If one now takes the customer perspective instead of the perspective of the single innovating body, the customer or buying firm can ask several innovating bodies to submit offers in form of utility-price combinations for a functionally specified product. Each innovating body or potential supplier can offer for him feasible utility-price combinations. If one or more innovating bodies would like to offer new or unforeseen product characteristics within given attributes or even new exclusive attributes, the customer has to take a decision to update the attribute catalogue or in other words his preference curve.

Following, a negotiation over price and technical utility can take place. To prevent from the temptation to use technical innovation from one supplier by updating the customer specification and to renegotiate only over price with the risk that the true innovating body will lose his innovative idea to a competitor, the negotiation in such case should run in parallel over technical utility and price.

If the given scenario occurs in a business-to-consumer environment, the reservation price per preferred stimulus of the customer is enough to decide about possible product-price points.

If the scenario happens in a business-to-business environment, an overarching target price can be applied. In such a case, utility-price combinations on or above the preference curve of the customer can be located below and left or above and right of the customer's customer target price. This situation creates a so far unobvious target area for the customer and the innovating body. The customer of the customer can be asked if he wants a relatively higher utility for a relatively lower price increase over the set target price. By this possibility, the consumer can be directly linked-in to a business-to-business decision.

This chapter develops the described model with the practical benefit to integrate a variable technology dimension into sourcing decisions. At the same time, the model theoretically prevents from the risk of wrong product-price allocations by introducing the active measurement of a technical utility value and a two-dimensional parallel sourcing over technology and price. The decisive element for the successful application of the model is the digitalization. Digitalization allows on one hand to effectively rank the product stimuli according to the customer's perspective preference and to calculate in real time the corresponding utility values. Further, digitalization allows capturing new and tracking updated supplier offers during the negotiation phase always translated to the corresponding utility-price combination. As a last step, digitalization allows the integration of customer target prices in real-time, leading to an effective decision process that involves in parallel preference utility-price values, feasibility utility-price values and the customer's customer target prices.

9.2 Why supplier innovation should be in the scope of purchasing

Several researchers discuss the assimilation of external knowledge as a necessity to minimum accelerate or to reinforce internal innovation (Azadegan and Dooley 2010; Chesborough 2003; Winter 2015). Regarding the innovation target itself, it was shown by Simon (2007) that if a company wants to reach market leadership, innovation and technological advantage are the key success factors. A similar finding was concluded in a study of the Fraunhofer Institute for Systems and Innovation Research about the structure and drivers of innovation success in the German machine building industry. The two elements innovation and technology and concurrent quality leadership were found to be the decisive points for the company's success (Kinkel and Som 2007). Looking into the buyer supplier relationship, after intense studies on the question of identifying and pulling innovation from supplier to customer (Dyer and Singh 1998; Koufterous et al. 2005; Song and Di Benedetto 2008) also the inverse direction of pushing innovation from supplier to customer was researched (Wagner and Bode 2014; Monczka et al. 2010; Schiele 2012).

The necessity of supplier innovation integration to the customer product and process development is seen especially true in turbulent economic environment and rapid changes in the development of new products (Noblet and Simon 2010; Phillips et al. 2006; Johnsen et al. 2006; Bessant et al. 2005; Zahra and George 2002). In an environment where the available product development time and own development resources decrease and the pressure on robust innovation increases at the same time, a significant contribution of innovation by suppliers is a necessary element for success (Clark 1989; Hartley 1997; Ragatz et al. 1997; Primo and Amundson 2002). Le Dain et al. (2011) explain in addition, that suppliers play a more important role in new product development with an increasing share of component outsourcing. Van Weele (2003) linked this aspect with the resource leverage that can be created if not only internal research and development resources are used, but if the customer is able to mobilise external capabilities in the new product development. On top, Cassiman and Veugelers (2002, 2006) argue that innovating firms are performing better when they combine internal innovation activities with external technology sourcing.

Schiele (2010a) shows that successful companies from an innovation standpoint intensively work with more than double the number of innovative suppliers than unsuccessful companies do. A prerequisite for successful innovation contribution from suppliers thereby is the early supplier involvement during the development activity on the one hand and the collaboration between purchasing and R&D department of the buying firm on the other hand (Schiele and Haas 2007; Schiele 2010b).

The key towards success in this context is excellent supplier management that gives the opportunity to gain competitive advantage (Johnson and Leenders 2010). Calvi (2012) introduces different perspectives of supplier innovation based on Brem and Tidd (2012), whereas Inemek and Matthyssens (2012) confirm that collaborative buyer supplier relations enhance innovation. At the same time they argue that cross-functional links foster supplier innovation, but collaborative communication is not enough for successful innovation integration. Only own design responsibility by the supplier combined with collaborative communication leads to a beneficial result. Schiele et al. (2010, 2011) go in the same direction, as they prove that a preferred customer status has a positive effect on supplier innovativeness, also prices are better if a preferred customer status is present. Schiele et al. (2012) introduce a model for the preferred customer by using a social exchange perspective to link customer attractiveness, supplier satisfaction and preferred customer status. This is an important study, as the supplier satisfaction plays a decisive role in innovation transfer relations especially if sustainability is required. In other words, the innovating body has to be rewarded for bringing up innovation to the customer. Diehl and Stroebe (1987) and Hirt et al. (2008) define mandatory elements for creativity with the creative element itself being first thinking by yourself, the creativity enhancement of good mood, self-confidence, and optimism. The aspect of necessary presence of optimism is again confirmed by Schiele et al. (2012) following on Thibaut and Kelley (1959) that first stated the assumption that actors will use not only absolute, but also relative criteria to evaluate the outcome of an exchange relationship. This implies that to eventually decide upon whether a relationship should be continued or not, actors are influenced by the availability of alternatives. Diehl and Stroebe (1987) continue to explain that environment and situation are decisive; hard work, expertise, broad interest, and openness are prerequisites for creative moves. In the international context, Harhoff et al. (2014) show that inner-cultural ties are positively influencing the transfer of innovation. From the perspective of the supplier or innovating body, Wagner and Bode (2014) analyze which aspects act as safeguards for the product or process innovation sharing started by the supplier. They find that contract length, relationship age and strong cooperation can protect the relationship-specific investment of the supplier into a customer in a way that the supplier is more willing to push innovation towards the customer. Wagner and Bode (2014) in this regard extend the observations of Henke and Zhang (2010), who come to the conclusion that those customers with which suppliers have the closest working relationship have a higher chance to get access to supplier innovation than other customers would have.

Regarding the measurement of innovation, some scholars look into the indirect management such as number of patents, research, and development investments relative to sales, top line improvement out of recent innovations or bibliometric data (Smith 2005). The technical or direct measurement is present in form of discussion on outputs, sources, instruments, innovation methods and supplier collaboration (Hansen 2001; Guellec and Pattinson 2001; Luzzini et al. 2015). Whereas within this direct measurement approach the object based approach in contrary to the subject or input based approach focuses on the objective output of the innovation process, on the technology itself (Smith 2005; Archibugi and Pianta 1996). Both approaches define innovation in the sense of Schumpeter (Röpke and Stiller 2006) as the market introduction of a new product or process. The subject based approach is somewhat also indirect, as it on the one hand measures a specific firm, but this on the other hand based on general inputs. Within the subjective approach, Evangelista et al. (1998) show that capital expenditure related to research and development investments is across all analyzed industry sectors the most powerful component for innovation indication. Evangelista et al. show that research and development therefore is embedded in capital, but needs more specific measurement to be explicitly detectable in a forward looking scenario. The objective approach however looks on the innovation itself. Here one example is the SPRU database, which was developed by the meanwhile fifty year old Science Policy Research Unit at the University of Sussex. The idea was to collect information on major technical innovations in the British Industry and to analyze those. The research unit used the help of around 400 technical experts to identify major innovations across all sectors of the British economy. The approach is a direct measurement, but also backward looking. Other databases have done similar approaches (Muldur 2003; Acs and Audretsch 1990). All of them have in common and mention as a drawback that they are not differentiating between incremental and radical changes and especially that the measurement of innovation is therefore not ad hoc or in real-time.

The presented model in this chapter combined with digitalization has the power to take this measurement problem out and allows a direct and ad hoc or real-time measurement of innovations. It therefore overcomes a blind spot in innovation research regarding the possibility to give each incremental innovation a value that is not equal to money. Digitalization helps to rank product value added out of the view of a customer even under participation of a multitude of decision makers and users to see innovation as a technical value. Combining the technical value with the core purchasing element of the reservation price, each technical value in the supplier view gets a price and in the customer view a cost. As a result, it is possible to negotiate transaction points over price and technical value. As digitalization not only allows more efficient multiparty communication but also creates new markets, digitalization is the key mean to run such a utility-price negotiation. In a market where innovation is primarily generated in business-to-business relationships before it hits the consumer in following business-to-consumer relationships, the created and visualized transactions points can be related to customer target prices. As a consequence, markets with different technology levels can be opened up.

9.3   Pitfalls in innovation management with suppliers

The target of the authors is to create a model that allows practical application, e.g. in an industrial environment. In this regard, the practical pitfalls and respective countermeasures have to be addressed as well when pursuing innovation management with suppliers.

The first practical problem or pitfall for the purchaser when using contribution from suppliers in innovation is that the suppliers normally have to be selected in an early, from the purchaser point of view too early, project phase. And as the same suppliers hardly can be changed in a later project phase, competition needs to be ensured before the collaboration phase between the selected supplier and the customer. As during that decision time the final product design logically is not fixed yet, the technological dimension needs to be added to the sourcing concept of the buying firm (Schupp 2004).

This aspect needs to be emphasized as many purchasing organizations look for lowest possible product prices, good quality, and high delivery reliability first, but they do not explicitly include technology or supplier innovation contribution to their primary target matrix (Schumacher et al. 2008). In this context, Glantschnig (1995) asked member companies of the German Association of Materials Management, Purchasing and Logistics what their main purchasing targets are. The answers were lowest possible price, a good quality, and a high level of delivery reliability. Grochla and Schönbohm (1981) in addition introduced the target of cost reduction. Dobler et al. (1995) list similar targets of the purchasing and logistics function. This target focus shows that purchasing organizations do not primarily strive for supplier contributed optimized designs and technological fit with the needs of the buying firm or the end customer. The purchasing organizations also do not necessarily work towards an optimized product-price allocation for items that they buy.

Schumacher et al. (2008) however, show in their study about the main leverages in purchasing that the product optimization has the biggest leverage on the bottom line of a company. Compared to this technical leverage, pooling of demand, global sourcing, and process improvement have significantly less contribution. A good relationship between customer and supplier however results in significant improvements to the bottom-line as well. As a result, the study suggests that the target of product optimization should be added as one important element to the targets of a purchasing organization.

Therefore, the authors of this chapter conclude that the technology dimension itself needs to be added to the sourcing concept of the buying firm.

A second pitfall is the wrong or missing determination of functional preferences related to respective reservation prices of the customer or target prices of the customer's customer. In a business-to-business environment, hereby the target price of a supply chain transaction primarily is determined by the customer of the buying firm.

The competition phase of a sourcing process can have one or multiple winners that in many cases will start with the product development process after the awarding. If the market rules do not allow significant re-negotiations with the customer of the buying firm after project awarding, especially in cases where a high degree of innovation is required, the probability for a single source supplier is high. At the same time the risk for future financial loss incurred by the winning supplier is equally high. The latter is widely described under the phenomenon of the winner's curse (Capen et al. 1971). Here the winning bidder in a First-Price-Sealed-Bid auction tends to overestimate the unknown value of the auction object or underestimates the cost and consequently bids too optimistically. As a consequence, the winner of the sourcing risks ending up with systematically limited margins or even with losses. Of course, this phenomenon only occurs with bidders that act irrationally, but in business practice this situation happens from time to time. Literature analyses two main influencing factors (Engel et al. 2006). The first one states that the higher the number of bidders, the higher the risk for the occurrence of the winner's curse is. This aspect is widely accepted to be true. The second factor is the degree of uncertainty. Here many authors agree with the statement that the probability and the extent of the winner's curse increases with the degree of uncertainty over the sourcing object (Kräkel 1992; Mehner 1999; Cox and Isaac 1984). Uncertainty for the supplier can arise from unclear technical aspects, from an undefined business volume or business environment or from changing expectations with regards to technological requirements.

As a recommendation for the buying firm's purchasing organization, this means that in case of absence of post-sourcing re-negotiation possibilities with the buying firm's customer, the determination of functional preferences together with different reservation prices for respective functional product characteristics is necessary in order to eliminate the risk of financial loss by wrong product-price allocations.

A third pitfall in innovation management with suppliers is to close the design at a too early stage. To reduce the risk of product failure, the buying firm tends wanting to specify product parameters in a very precise way from the beginning of the project. If the buying firm does that, three types of risk can occur. The first one is a missed chance for innovation in an unexpected area. This could for example happen if a not awarded supplier could have reached superior product-price combinations in comparison with the winning supplier, but simply could not show his capabilities because of restricted product definitions in the beginning. Another risk could be that the selected supplier turns out to have technological disadvantage or even performance problems to achieve the specified product parameters if he is out of his technological sweet spot. A further risk is a non-optimized product-price combination which could lead to an overshooting of previously defined target cost or a wrong product-price allocation. The winning supplier could have been in a position to offer a better product for the same price or a less advanced product with an over-proportionally lower price. All such risks result out of ignored or simply unused supplier core or best-cost competencies in product or process innovation. Therefore, it is evident for the buying firm to keep their own design open and to start with a functional specification rather than with a finalized design both on customer product and on sourcing object level.

A forth pitfall when trying to sustainably buy innovation is the rewarding problem. As the potential suppliers have to submit new technologies, own technical solutions, and innovations, they must be rewarded for innovations that are used in final designs instead of the buying firm using and internalizing the innovation and sourcing it at another competing supplier. Therefore, the sourcing process has to run simultaneously over the technology dimension and the price dimension in order to avoid that the buying firm's preference is updated with the different supplier innovation contributions before the final sourcing. Otherwise, the rewarding of innovative suppliers is endangered. Therefore, it is recommended to run an innovation sourcing process as a parallel process over the two dimensions price *and* technology.

9.4   Purchasing model for sourcing of utility-price combinations

The idea behind the present purchasing model is to create a solution space of possible product-price combinations instead of focussing on a fixed set of functional product characteristics or a fixed specification. In this way, the potential suppliers have the chance to introduce different technical and functional solutions that lead to several different product-price combinations. To guide the reader through the model, a reference to a practical example is used. In the example the buying firm intends to buy a display for a rear seat entertainment system to be sold to a customer that is producing passenger cars.

The model is based on a three-step concept. In the first step, the user is defining his preferences. The second step creates the possible product-price combinations offered by the potential suppliers and the feasibility curve. In the third step, the feasible product-price combinations will be put in context with the preferences of the buying firm. Orientated at the value added of the product-price combination offered in relation to the preference curve and under the condition of the customer target price, an optimal supplier decision can be taken. By definition, the term 'product' is used in the sense of the component or service to be purchased from the suppliers.

9.4.1 A developer's survey reveals most important product attributes and characteristics

The development of the preference curve starts with a survey amongst the buying firm's key development and marketing staff identifying the relevant attributes of the targeted product or components to be purchased (Table 9.1).

Table 9.1 The buying firm sets relevant attributes and characteristics

Ranking	Attributes		Characteristics	
1.	A1	Display size [inch]	A1.1	7
			A1.2	10
			A1.3	12
2.	A2	Display weight [gram]	A2.1	800
			A2.2	500
			A2.3	300
3.	A3	Electricity consumption [watt]	A3.1	10
			A3.2	6
			A3.3	4
4.	A4	Response time [ms]	A4.1	3
			A4.2	2
			A4.3	1

For each selected attribute possible characteristics have to be defined. In the model it is suggested to use 4 attributes and 3 characteristics for each attribute. The possible number of combinations equals to the number of characteristics to the power of the number of attributes. Therefore, maximum 81 combinations are possible in this model (Backhaus et al. 2008). Of course, one can choose to increase or decrease the number of attributes and characteristics. However, the chosen number of 4 and 3 gives enough possibilities and at the same time limits the complexity to a practical level. Another argument for limiting the number of characteristics is that during the evaluation of the preferred and the feasible product-price combinations, an interpolation between the different combinations will be done. In the chosen example the attributes display size, display weight, electricity consumption, and response time are used to characterize the purchased product.

9.4.2 Ranking of the top 10 product stimuli

In the next step, the buying firm is requested to rank the best 10 product stimuli (Table 9.2). Best in the sense of the model means the preferred combinations. Each stimulus consists of one characteristic per attribute.

In the example, the best combination is a 12 in. display with 300 g weight, 4 W electricity consumption and 1 ms response time. Key in this development step of the preference curve is to assign a reservation price to each stimulus. The reservation price is the maximum price that the buying firm is willing to pay for the respective

Table 9.2 Ranking of stimuli according to preference and likelihood of acceptability

Ranking		Characteristics for Attribute				Reservation price
	A1	A2	A3	A4		
1.	12	300	4	1		100
2.	12	500	4	2		98
3.	12	500	4	3		93
4.	12	300	6	2		85
5.	12	500	6	2		83
6.	12	500	6	3		80
7.	10	300	4	2		79
8.	10	500	4	3		76
9.	12	800	10	3		74
10.	10	500	6	3		73

stimulus. As the stimuli are ranked, the highest reservation price will be found at the top ranked stimulus. Table 9.2 lists 10 stimuli for good visibility. Of course, theoretically all 81 stimuli would be listed, but for the purpose of easier handling only the first 10 stimuli are used in the model going forward.

The setting of attributes and characteristics and also the ranking of the stimuli in business practice is only efficient with the help of digitalization. One or more stakeholders, mostly development engineers and marketing specialists, can define attributes with characteristics and rank them via a web-based tool. Out of the product planning or controlling group of the buying firm, the respective reservation prices can be added.

9.4.3 Conjoint analysis of stimuli and transformation into standard utilities

Step three in the development of the preference curve is the conjoint analysis itself of the stimuli (Table 9.3). The method of the conjoint analysis identifies the importance of each attribute and characteristic (Luce and Tukey 1964; Backhaus et al. 2008).

In the conjoint analysis, each attribute will receive an importance in percent totalling to 100 percent. Within each attribute, for the defined characteristics a

Table 9.3 Calculation of the utility value by conjoint analysis

Attributes		Importance in %	Characteristics		Standard utility
A1	Display size [inch]	50	A1.1	7	0.000
			A1.2	10	0.200
			A1.3	12	0.500
A2	Display weight [gram]	25	A2.1	800	0.000
			A2.2	500	0.200
			A2.3	300	0.250
A3	Electricity consumption [watt]	20	A3.1	10	0.000
			A3.2	6	0.050
			A3.3	4	0.200
A4	Response time [ms]	5	A4.1	3	0.000
			A4.2	2	0.030
			A4.3	1	0.050
	Σ	100			

utility value is calculated. By normalizing the values, the least important characteristic within an attribute gets assigned to the standard utility of 0. The highest valued characteristic gets assigned to the weight of the complete corresponding attribute. Needless to say, that also this step profits from digitalization over a web-based tool.

9.4.4 Ranking of the best 10 stimuli by utility and reservation price

In step four, the standard utilities out of step three are used to calculate the utility of each stimulus. The calculation is simply adding the standard utilities of each characteristic within the 10 selected product stimuli (Table 9.4). By the method of normalization the maximum value of a stimulus utility is 1. The minimum is 0.

In step five of the preference curve creation, the reservation price for each stimulus and the utility of the same stimulus are put in context by entering both in a graph (Table 9.5). The reservation price is put down on the x-axis, the utility

Table 9.4 Referencing of reservation price and utility of the 10 best stimuli

Ranking	1.	2.	3.	4.	5.	6.	7.	8.	9.	10.
Utility stimulus	1.000	0.930	0.900	0.830	0.780	0.750	0.680	0.600	0.500	0.450
Reservation price	100	98	93	85	83	80	79	76	74	73

Table 9.5 Matching of utility and reservation prices to the preference curve

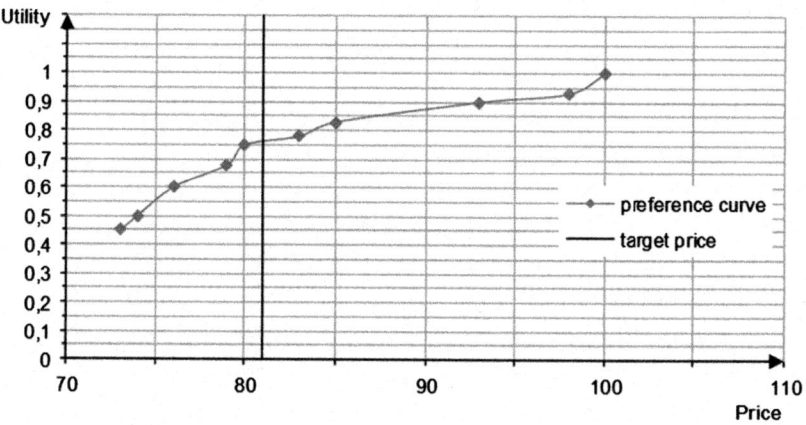

on the y-axis. The final selection for a product-price combination in business practice does not only depend on the reservation price of the buying firm, but also on the target price of the buying firm's customer. This customer target price is introduced to the model at this point.

Of course, it can happen that some acceptable stimuli from a buying firm's point of view have higher reservation prices compared to the customer target price. This situation will be discussed further after the development of the feasibility curve with regards to offered product-price combinations by the potential suppliers.

9.4.5 Development of the feasibility curve based on supplier innovation

Based on an open functional specification, the potential suppliers are asked to offer their product-price solutions to the buying firm. In a concept competition, each potential supplier is asked to offer several possible product solutions that also can have different prices.

The first step for the development of the feasibility curve uses the utility values of the characteristics defined in the development of the customer preference curve before (Table 9.6).

If suppliers offer characteristics that do not match with the selected characteristics by the buying firm, the model will interpolate the utility values of the offered and the selected characteristics. The utility values of each characteristic

Table 9.6 Supplier offers translated in utility values and related offered price

Offer	Supplier	Utility A1	Utility A2	Utility A3	Utility A4	Σ Utility	Price
1	A	0.400	0.200	0.200	0.050	0.850	105
2	C	0.400	0.150	0.200	0.030	0.780	100
3	B	0.300	0.250	0.200	0.020	0.770	94
4	A	0.550	0.200	0.000	0.000	0.750	85
5	A	0.350	0.250	0.050	0.050	0.700	83
6	B	0.400	0.150	0.100	0.030	0.680	80
7	C	0.200	0.350	0.100	0.000	0.650	79
8	B	0.500	0.100	0.000	0.000	0.600	76
9	C	0.200	0.100	0.000	0.050	0.350	75
10	B	0.100	0.100	0.000	0.030	0.230	72

are added to a sum of utilities. In addition to the utility sum of the offered stimuli by the suppliers, the corresponding offering price is included in the model. As mentioned before, suppliers might offer new product attributes that are decision relevant. If so, the steps 1 to 5 of the model have to be repeated.

9.4.6 Utility-price ranking of the 10 best bids

The next step of the model is the negotiation between the buying firm and the competing suppliers.

Different to conventional negotiation situations, the suppliers can either offer better prices or better utilities. The latter can be done by offering better characteristics compared to the previous solution. This process can also lead to a situation where the utility value will increase and the price will increase as well. In the example (Table 9.7) this situation is shown by supplier B who is offering 100 € and a 0.930 utility in offer number 3 compared to 94 € and 0.900 utility in offer number 2. The format displayed allows three rounds of negotiation. Nevertheless, more rounds are possible in real business situations.

For each negotiation round a feedback is required. The feedback indicates if the offered product-price combination is above or below the preference curve of the buying firm. If the offered product-price combination is above the preference curve, the feedback is 'g' for green. The meaning is that the product-price point

Table 9.7 Three offer rounds allow price and utility changes

Supplier	Offer 1		Feedback	Offer 2		Feedback	Offer 3		
	Utility	Price		Utility	Price		Utility	Price	Feedback
A	0.850	105	r	0.880	105	r	0.950	105	r
B	0.780	100	r	0.900	94	r	0.930	100	r
C	0.770	94	r	0.900	95	r	0.900	94	r
A	0.750	85	r	0.850	84	g	0.890	85	g
B	0.700	83	r	0.830	84	g	0.860	83	g
A	0.680	80	r	0.740	78	g	0.820	80	g
C	0.650	79	r	0.680	77	g	0.800	79	g
B	0.600	76	g	0.700	79	g	0.700	76	g
C	0.350	75	r	0.400	73	r	0.400	73	r
B	0.230	72	r	0.350	75	r	0.330	72	r

is inside the area of acceptable solutions for the buyer. The feedback 'r' for red indicates that the offered solution is not acceptable for the buyer as it is below his preference curve.

In this element of the model, digital feedback to the suppliers allows to start immediate improvement actions by the suppliers to put themselves in a better position in the following round of negotiation.

9.4.7 Winner determination by comparing for preference and feasibility curve

In the following step, the winning combinations are listed. Winning combinations are defined as the combinations after the final negotiation round with the feedback green.

If the result of the negotiation has more than one winning combination, another two steps to determine the best solution have to be taken.

First, the offered prices of the winning combinations have to be mirrored against the buying firm's customer target price. In general, it can be assumed that an acceptable solution for the customer has to beat his target price. Nevertheless, there might be situations where a customer would be willing to discuss an attractive utility-price combination that is even above his target price. Such a situation could occur if e.g. a customer has a product placing strategy for low end and high end market segments at the same time.

Secondly, the question needs to be answered which one of the winning combinations is the best combination out of the buyer's point of view.

In the example 3, winning combinations have a utility of 0.8 at a price of 79 €, a utility of 0.7 at a price of 76 € and a utility of 0.82 at a price of 80 €. If the buyer would choose the winning combination with the lowest price, he would decide for a utility of 0.7 at a price of 76 €. But as the target of the model is to identify the best offer based on price *and* utility, the buyer could choose the winning combination with the highest delta utility towards the preference curve. With this argument, the combination with the utility of 0.8 and the price of 79 € would be the best combination. In order to identify such kind of best utility-price combinations, all feasible solutions above the preference curve are listed and ranked by their delta utility to the preference curve. Table 9.8 shows which combinations are below or above the customer target price. In Table 9.9 the presented example is aggregated into the preference curve and a feasibility curve before and after negotiation in one graph.

Table 9.8 Identification of the utility delta between feasibility and preference curve

Supplier	Price	Required utility	Offered utility	A (offered—required) Utility	Above/below target price
C	79	0.680	0.800	+0.120	Below
B	76	0.600	0.700	+0.100	Below
B	83	0.780	0.860	+0.080	Above
A	80	0.750	0.820	+0.070	Below
A	85	0.830	0.890	+0.040	Above

Table 9.9 Feasibility curve of final offers in context with target price and preference curve

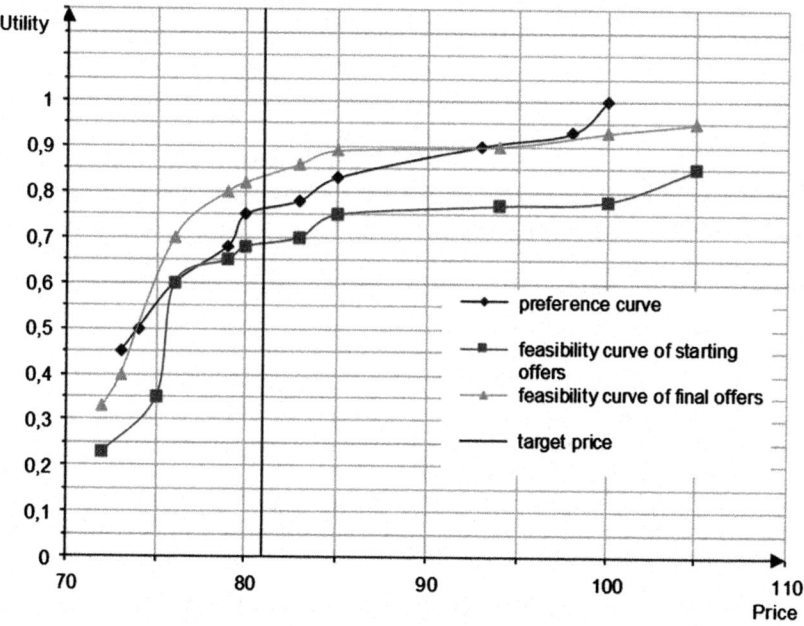

By the introduction of the target price line of the buying firm's customer, the user of the model can identify the relevant winning solutions. As mentioned above, also the two other green utility-price combinations are element of the solution space of the model. They are above the target price line of the customer's customer, but represent high enough utilities. Therefore, the customer might offer an upgrade product to the customer's customer that he could sell to a higher end market.

9.5   Conclusion and future research directions

The present model shows how to effectively include the technology dimension into sourcing decisions where a supplier contribution with regards to innovation is required. The model is created for the use in business practice supported by digitalization. It allows systematic early supplier involvement without losing the competition phase and can be applied for innovation management with suppliers in different sectors and used for various products. A transformation from product to component level is possible, as in such cases the product innovation would be gradually replaced by process innovation. In the next development step for the effective application, digital buying portals or buyer-supplier platforms have to be enhanced, so the buying firm can process the technology dimension by entering attributes, characteristics to get to product stimuli and to get to calculated corresponding utilities. Invited suppliers can set their bids in form of utility-price combinations and get real-time feedback about their ranking in the negotiation. If the customer of the customer can be linked to the portal as well, the target prices of the customer's customer can be used in order to allow real-time decision making that is including even the solution space above the customer's customer target price. Both elements allow achieving the targeted time pressure in sourcing decisions that enhances inherent competition (Arnold and Schnabel 2008).

References

Acs, Z., Audretsch, D.: Innovation and Small Firms. MIT Press, Cambridge, Mass (1990)

Archibugi, D., Pianta, M.: Measuring technological change through patents and innovation surveys, Technovation 16(9), 451–468 (1996)

Arnold, U., Schnabel, M.: Economic effects of electronic reverse auctions: a procurement process perspective. In: Parente, D.H. (eds.) Best Practices for Procurement Auctions. Information Science Reference, pp. 57–76. Hershey, New York (2008)

Azadegan, A., Dooley, K.J.: Supplier innovativeness, organizational learning styles and manufacturer performance: an empirical assessment. J. Oper. Manag. 28(6), 488–505 (2010)

Backhaus, K., Erichson, B., Plinke, W., Weiber, R.: Multivariate Analysemethoden, eine anwendungsorientierte Einführung, 12. edn., pp. 557–618. Springer, Berlin (2008)

Bessant, J., Lamming, R., Noke, H., Phillips, W.: Managing innovation beyond the steady state. Technovation 25, 1366–1376 (2005)

Brem, A., Tidd, J. (eds.): Perspectives on Supplier Innovation: Theories Concepts and Empirical Insights on Open Innovation and the Integration of Suppliers. Imperial College Press, UK (2012)

Calvi, R.: Book review: perspectives on supplier innovation: theories concepts and empirical insights on open innovation and the integration of suppliers. In: Brem, A., Tidd J. (eds.). Imperial College Press, UK (2012), In: J. Purchasing Suppl. Manag. 18, pp. 282–283 (2012)

Capen, E.C., Clapp, R.V., Campbell, W.M.: Competitive bidding in high-risk situations. J. Petrol. Technol. **23**, 641–653 (1971)

Cassiman, B., Veugelers, R.: R&D cooperation and spillovers: some empirical evidence from Belgium. Am. Econ. Rev. **92**(4), 1169–1184 (2002)

Cassiman, B., Veugelers, R.: In search of complementarity in innovation strategy: internal R&D and external knowledge acquisition. Manag. Sci. **52**(1), 68–82 (2006)

Chesborough, H.: Open innovation: the new imperative for creating and profiting from technology. Harvard Business School Press, Boston (2003)

Clark, K.B.: Project scope and project performance: the effects of parts strategy and supplier involvement on product development. Manag. Sci. **35**(10), 1.247–1.263 (1989)

Cox, J.C., Isaac, R.M.: In search of the winner's curse. Econ. Inq. **22**(4), 579–592 (1984)

Diehl, M., Stroebe, W.: Productivity loss in brainstorming groups: toward the solution of a riddle. J. Pers. Soc. Psychol. **53**(3), S. 497–509 (1987)

Dobler, D.W., Lee, L., Burt, D.N.: Purchasing and Supply Management, 6. Aufl. McGraw-Hill, New York (1995)

Dyer, J.H., Singh, H.: The relational view: cooperative strategy and sources of interorganizational competitive advantage. Acad. Manag. Rev. **23**(4), 660–679 (1998)

Engel, A., Hauk, E., Ganuza, J.J., Wambach. A.: Risk management in procurement auctions. In: Dimitri, N., Piga, G., Spagnolo, G. (eds.) The Handbook of Procurement, Cambridge University Press, Cambridge, chapter 13 (2006)

Evangelista, R., Sandven, T., Sirilli, G., Smith, K.: Measuring innovation in European industry. Int. J. Econ. Bus. **5**(3), 311–333 (1998)

Glantschnig, E.: Industrieller Einkauf heute. Beschaffung aktuell **42**(12), 25–29 (1995)

Grochla, E., Schönbohm, P.: Beschaffung in der Unternehmung: Einführung in eine umfassende Beschaffungslehre. Poeschel, Stuttgart (1981)

Guellec, D., Pattinson, B.: Innovation surveys: lessons from GECD Countries' experience. STI Rev. **27**, 77–102 (2001)

Hansen, J.A.: Technological innovation indicators: a survey of historical development and current practice. In: Feldmann, M.P., Link, A. (eds.) Innovation Policy in the Knowledge-Based Economy, pp. 73–103. Kluwer, Dordrecht (2001)

Hartley, J.L.: Suppliers' contribution to product development: an exploratory study. IEEE Trans. Eng. Manag. **44**(3), 258–267 (1997)

Harhoff, D., Mueller, E., Van Reenen, J.: What are the channels for technology sourcing? Panel data evidence from German companies. J. Econ. Manag. Strateg. **23**(1), 204–224 (2014)

Henke Jr., J.W., Zhang, C.: Increasing supplier-driven innovation. Sloan Manag. Rev. **51**(2), 41–46 (2010)

Hirt, E., Devers, E.E., McCrea, S.M.: I want to be creative: exploring the role of hedonic contingency theory in the positive mood-cognitive flexibility link. J. Pers. Soc. Psychol. **94**, 214–230 (2008)

Inemek, A., Matthyssens, P.: The impact of buyer-supplier relationships on supplier innovativeness: an empirical study in cross-border supply networks. Ind. Mark. Manag. 1–15 (2012)

Johnsen, T., Phillips, W., Caldwell, N., Lewis, M.: Centrality of customer and supplier interaction in innovation. J. Bus. Res. **59**, 671–678 (2006)

Johnson, F.P., Leenders, M.R.: Supplier Relationship Management. In: Hallikas, J., Kähkönen, A.-K., Lintukangas, K., Virolainen, V. M. (eds.) Proceedings of the 19th Annual IPSERA conference in Lappeenranta 2010, pp. 212–221. Lappeenranta, Finland (2010)

Kinkel, S., Som, O.: Strukturen und Treiber des Innovationserfolges im Deutschen Maschinenbau. Fraunhofer-Institut für System- und Innovationsforschung ISI, Karlsruhe, **41**(3) (2007)

Koufteros, X., Vonderembse, M., Jayaram, J.: Internal and external integration for product development: the contingency effects of uncertainty, equivocality, and platform strategy. Decis. Sci. **36**(1), 97–133 (2005)

Kräkel, M.: Auktionstheorie und interne Organisation. Gabler, Wiesbaden (1992)

Le Dain, M.-A., Calvi, R., Cheriti, S.: Measuring supplier performance in collaborative design: proposition of a framework. R&D Manag. **41**(1), 61–79 (2011)

Luce, R.D., Tukey, J.W.: Simultaneous conjoint measurement. J. Math. Psychol. **1**(1), 1–27 (1964)

Luzzini, D., Amann, M., Caniato, F., Essig, M., Rochi, S.: The path of innovation: purchasing and supplier involvement into new product development. Ind. Mark. Manag. **47**, 109–120 (2015)

Mehner, M.: Winner's curse. TU Chemnitz, Grin (1999)

Monczka, R.M., Carter, P.L., Scannell, T.V., Carter, J.R.: Implementing Supplier Innovation: Case Study Findings. CAPS Research, Tempe, AZ (2010)

Muldur, U. (ed.), European Commission, 3rd European Report on Science and Technology Indicators, EUR20025, Luxembourg (2003)

Noblet, J.-P., Simon, E., La capacité d'absorption, un état de l'art, in: Management&avenir, **5**(35), pp. 33–50 (2010)

Phillips, W.E., Noke, H., Bessant, J., Lamming, R.: Beyond the steady state: managing discontinuous product and process innovation. Int. J. Innov. Manag. **10**(2), 1–23 (2006)

Primo, M.A.M., Amundson, S.D.: An exploratory study of the effects of supplier relationships on new product development outcomes. J. Oper. Manag. **20**, 33–52 (2002)

Ragatz, G.L., Handfield, R.B., Scannell, T.V.: Success factors for integrating suppliers into new product development. J. Prod. Innov. Manag. **14**(3), 190–203 (1997)

Röpke, J., Stiller, O. Schumpeter, J. (eds.): Theorie der wirtschaftlichen Entwicklung: Nachdruck der 1. Auflage von 1912 (2006)

Schiele, H.: Innovationen von und mit Lieferanten—Ergebnisse einer quantitativen Studie. BME-Report, Enschede (2010a)

Schiele, H.: Early supplier integration: the dual role of purchasing in new product development. R&D Manag. **40**(2), 138–153 (2010b)

Schiele, H.: Accessing supplier innovation by being their preferred customer. Res. Technol. Manag. **55**(1), 44–50 (2012)

Schiele, H., Calvi, R., Gibbert, M.: Customer attractiveness, supplier satisfaction and preferred customer status: Introduction, definitions and an overarching framework. Ind. Mark. Manag. **41**(8), 1178–1185 (2012)

Schiele, H., Haas, C.: Innovationsquelle Lieferant. Beschaffung aktuell **38**(4), 30–31 (2007)

Schiele, H., Veldman, J., Hüttinger, L.: Supplier innovativeness and supplier pricing: the role of preferred customer status. In: Hallikas, J., Kähkönen, A.-K., Lintukangas, K., Virolainen, V. M. (eds.) Proceedings of the 19th Annual IPSERA conference in Lappeenranta 2010, pp. 378–394. Lappeenranta, Finland (2010)

Schiele, H., Veldman, J., Hüttinger, L.: Supplier innovativeness and supplier pricing: The role of preferred customer status. Int. J. Innov. Manag. **15**(1), 1–27 (2011)

Schiele, H., Veldman, J., Hüttinger, L., Pulles, N.: Towards a social exchange theory perspective on preferred customership—Concept and practice. In: Bogaschwesky, R., Eßig, M., Lasch, R., Stölzle, W. (eds.) Supply management research. Aktuelle Forschungsergebnisse 2012, pp. 135–151 (2012)

Schumacher, S.C., Schiele, H., Contzen, M., Zachau, T.: Die drei Faktoren des Einkaufs. Wiley-Vch, Weinheim (2008)

Schupp, F.: Versorgungsstrategien in der Logistik, Konzeption eines modularen Entscheidungsmodells. Gabler, Wiesbaden (2004)

Simon, H.: Hidden Champions des 21 Jahrhunderts. Campus, Frankfurt (2007)

Smith, K.: Measuring innovation. In: The Oxford Handbook of Innovation, pp. 148–177. Oxford University Press, New York (2005)

Song, M., Di Benedetto, C.A.: Supplier's involvement and success of radical new product development in new ventures. J. Oper. Manag. 26(1), 1–22 (2008)

Thibaut, J.W., Kelley, H.H.: The Social Psychology Of Groups. Wiley, New York (1959)

Van Weele, A.J.: Supply resource management: mobilising external capabilities. Executive Outlook 3(4), 82–92 (2003)

Wagner, S.M., Bode, C.: Supplier relationship-specific investments and the role of safeguardsfor supplier innovation sharing. J. Oper. Manag. 32, 65–78 (2014)

Winter, S.: Management von Lieferanteninnovationen: Eine gestaltungsorientierte Untersuchung über das Einbringen und die Bewertung. In: Bogaschewsky, R., Eßig, M., Lasch, R., Stölzle, W. (eds.) Supply Management Research, Advanced Studies in Supply Management, pp. 59–77 (2015)

Zahra, S.A., George, G.: Absorptive capacity: a review, reconceptualization, and extension. Acad. Manag. Rev. 27(2), 185–203 (2002)

About the Authors

Dr.-Ing. Florian Schupp is Senior Vice President Purchasing Automotive and Automotive Aftermarket of Schaeffler Group. He completed his PhD at the Technical University of Berlin in the field of strategy development in purchasing and logistics and has 19 years of purchasing experience in the companies Schaeffler, Continental, Siemens and SONY. He integrates practical purchasing and supply management work with academic research together with the University of Lappeenranta, Finland and the University of Catania, Italy. He teaches International Procurement at the Technical University of Berlin and Purchasing and Supply Management at Jacobs University in Bremen. Main research fields are purchasing strategy, behavioral aspects in purchasing, integration of supplier innovation and parametric auctions, working capital management, buyer-supplier relationship management and supply management. Dr.-Ing. Schupp is member of the Advisory Board of BLG Logistics, Bremen.

Matthias Rehm is general manager of Max Hauser, an international company specializing in the development and manufacturing of surgical instruments. Prior, he was plant purchasing manager in a large automotive supplier. Since joining as a PhD candidate at the Department of Management of the University of Antwerp, he researches and publishes in the fields of purchasing, open innovation, early supplier involvement, and buyer-supplier relationship management.

Opportunities in Emerging Markets – Purchasing with Start-ups

10

Florian Schupp, Heiko Wöhner und Alisée de Tonnac

Zusammenfassung

In diesem Interview berichtet Alisée de Tonnac über Chancen für und durch Start-ups in Emerging Markets. Alisée de Tonnac ist CEO & Co-Gründerin von Seedstar World, einem globalen Start-up-Wettbewerb, der das Leben von Menschen in Emerging Markets durch Technologie und Entrepreneurship positiv beeinflusst. Das Wichtigste für Start-ups seien die Kunden – als Konsumenten im B2C-Geschäft, als Einkäufer im B2B-Geschäft. Die erfolgreichen Start-ups schöpften aus der Fähigkeit, Innovationen zu entwickeln und schneller in marktfähige Produkte zu verwandeln als es in traditionellen Industrien möglich sei. Gerade in den Emerging Markets gäbe es enorme Potentiale, weil einige Industrien wie Banken für einen Großteil der Bevölkerung nicht erreichbar seien und Start-ups dieses Vakuum nutzen könnten.

F. Schupp (✉)
Bühl, Deutschland
E-Mail: schupp-florian@t-online.de

H. Wöhner
Baden-Baden, Deutschland
URL: http://www.linkedin.com/in/woehner/

A. de Tonnac
Genf, Schweiz
URL: http://www.aliseedetonnac.com

© Springer Fachmedien Wiesbaden GmbH 2018
F. Schupp und H. Wöhner (Hrsg.), *Digitalisierung im Einkauf*,
https://doi.org/10.1007/978-3-658-16909-1_10

10.1 Einleitung

Digitalisierung im Einkauf wird häufig zunächst mit Effizienzsteigerungen durch transparentere und verschwendungsärmere Vergabe- und Lieferprozesse in Verbindung gebracht. Es gibt aber auch indirekte und weit greifendere Veränderungen zum Beispiel durch die Neugestaltung von Märkten. Der technische Fortschritt in der Informations- und Kommunikationstechnologie hatte ab den 1990er Jahren den Produktfokus um die ergänzenden, zunehmend digitalen Services erweitert. Spätestens seit der Jahrtausendwende sind unzählige neuartige, rein digitale Geschäftsideen entstanden. Eine auf digitale Produkte und Services beschränkte Unternehmenstätigkeit verringert die Markteintrittsbarrieren für neue Unternehmen durch erheblich verringerte Investitionsbedarfe. Startup-Unternehmen können deshalb mit in traditionellen Industrien unbekannten Freiheiten in kurzer Zeit Produkte entwickeln, testen und verbessern. Start-ups verändern damit die Beschaffungsmärkte für Einkäufer.

Einerseits sind digitale Revolutionen, wie beispielsweise durch Airbnb in der Tourismusbranche, eine Bedrohung für die traditionellen Unternehmen und deren Einkaufsstrategie. Denn die Geschäftsmodelle zahlreicher Start-ups zielen auf die direkte Verbindung von Herstellern und Konsumenten durch das eigene digitale Produkt ab. Das Überspringen von Zwischenhändlern ermöglicht dabei attraktive Endkundenpreise und zugleich Margen für die Start-ups, die traditionellen Marktteilnehmern aufgrund der gegebenen Strukturen nicht ausreichen. Andererseits eröffnen Start-ups Potenziale für Einkäufer. Denn mit der weltweiten Digitalisierung sind bisher durch logistische Faktoren benachteiligte Beschaffungsmärkte, zum Beispiel in Afrika, erreichbar und offerieren Produktinnovationen junger, lokaler Unternehmen. Mit Ergänzung der digitalen Start-up-Produkte durch neue, passende Hardware entsteht zusätzlich ein Einkaufsbedarf in Unternehmen, für den diese bisher keine Expertise besitzen. Einkauf als Service könnte sich im wachsenden Markt entwickeln. Digitalisierung ist disruptiv für bestehende Märkte und schafft neue Möglichkeiten.

Und Digitalisierung ist auch disruptiv für Lebensläufe. Sie schafft neue Möglichkeiten, zum Beispiel im Lebenslauf von Alisée de Tonnac. Aufgewachsen in Singapur, Kalifornien und Frankreich, studierte Alisée de Tonnac in der Schweiz und Italien und begann als Produktmanagerin bei L'Oréal eine scheinbar vorgezeichnete Konzernkarriere. Mit 23 Jahren entschloss sich Alisée de Tonnac dann, ihren gut angesehenen Job bei L'Oréal zu kündigen und das bisher Erreichte gegen eine ungewisse Zukunft in der Welt der Start-ups, vor allem digitaler Start-ups, zu tauschen. Mit der Seedstars World ist sie heute weltweit in über 60 Städten aktiv. Ihr unternehmerisches und persönliches Engagement wurde bereits mehrfach ausgezeichnet. Beispielsweise zählt Forbes Alisée de Tonnac zu den „30 under 30 – Europe – Social Entrepreneurs".

Als wir in der Recherche für dieses Buch von Alisée de Tonnac gelesen haben und uns über ihr tägliches Engagement für Start-ups in entlegen scheinenden Regionen informierten, ist der Wunsch gereift, in einem Interview Näheres zu erfahren. Zum Glück hat Alisée de Tonnac zugesagt und uns gleich zwei Interviews ermöglicht: ein Mal digital irgendwo zwischen Nigeria und Lausanne, und ein Mal ganz persönlich im Co-Working-Space von Seedstars, einer dem Abbruch für Investmentprojekte geweihten Villa in Genf. Denn trotz aller Digitalisierung, so hat uns auch Alisée de Tannac erzählt, kommt es letztendlich auf den persönlichen Kontakt zu Menschen, auf das persönliche Treffen von Start-ups und Investoren, von Start-ups und Einkäufern an.

Im Interview berichtet Alisée de Tonnac, dass das Wichtigste für Start-ups die Kunden seien – als Konsumenten im B2C-Geschäft, als Einkäufer im B2B-Geschäft. Die erfolgreichen Start-ups schöpften aus der Fähigkeit, Innovationen zu entwickeln und schneller in marktfähige Produkte zu verwandeln, als es in traditionellen Industrien möglich sei. Gerade in den Emerging Markets gäbe es enorme Potenziale, weil einige Industrien wie Banken für einen Großteil der Bevölkerung nicht erreichbar seien und Start-ups dieses Vakuum nutzen könnten. Mit ihren Neugründungen entwickelt, fördert und vermarktet Alisée de Tonnac weltweit zahlreiche Unternehmen und Talente. Sie erklärte uns sogar, wie es möglich sei, Entrepreneure in geförderten Unternehmen anzustellen und zu motivieren. Mit ihrem einzigartigen Einblick in die Start-up-Szene gibt sie auch Tipps, wie traditionelle Unternehmen die Brücke zwischen ihrer jeweiligen Welt und den Start-ups erfolgreich schlagen können.

Das Interview haben wir in englischer Sprache geführt und veröffentlichen es auf den folgenden Seiten entsprechend.

10.2   Interview Alisée de Tonnac

Seedstars – Impacting people's lives in emerging markets.
Seedstars celebrates disruption, and the pioneers who create it. We roam the globe to connect with the digital changemakers of the emerging world shaping a new wave of social change. Our teams tread on the entrepreneurial arena across five continents, breaking stories about startups set upon defying the status quo. Our target represents 85% of human population, the opportunities encompass billions of lives, the scope of impact is infinite.

As an incubator of educators and investors, we sit at the forefront of innovation in emerging markets and we make it our mission to share these developments with you across the digital landscape.

We invest in people, we connect teams, we encourage risk. We empower entrepreneurial spirits to swim against the tide. Join us online or in person at our events anywhere around the globe. Join us as we seek to impact billions of lives in emerging markets.

URL: www.seedstars.com

Florian Schupp: *Dear Alisée, in preparation of this book the idea was also to look into purchasing markets that are rather unknown to most European purchasers because in these markets there might be opportunities evolving out of the digitalisation mega trend. Many industrial and service companies are active in traditional markets and some so called „new markets", which might be old markets for you such as China, India, Russia, Romania, and Slovakia. Purchasers know how to find suppliers and how to best work with them. But how can purchasers go beyond the known markets and get in contact with companies in for example Africa or in truly emerging markets? Could digitalisation help to get start-up companies easier in contact with companies which have demand and are looking for purchasing opportunities and good ideas in remote places? Is digitalisation an enabler?*

Alisée de Tonnac: In the question „will digitalisation influence the link between big corporations and start-ups", the answer is absolutely yes and you are seeing it not only in Europe but also in emerging markets. More and more, the investors, supporters, sponsors, and facilitators for start-ups in many markets are the Corporates.

What we have seen in the past is that corporates have failed in trying to innovate internally. For a while we thought R&D was the only way to innovate and indeed corporates had a tendency to invest a lot but with digitalisation the capacity of disruptive innovation has been shifted towards the flourishing start-up ecosystems where, more and more, any „wannabe entrepreneur" can launch her/his company (due to the democratisation of the costs related to building a company). What we have witnessed during our discussions with Corporates is that they thought that the disruption could come from in-house but in reality, in many ways, the Corporate DNA, the Corporate environment is „antagonist" to that of the DNA and environment of a start-up. Think of agility, very little bureaucracy, of high-risk takers always at the limit (or far of) of what is legal or not. Start-ups also hack their way, without standardised processes, whereas big Corporates are afraid of their branding, position, and public image and in many ways rightly so. But with the 4th industrial revolution and the acceleration at which new technologies and habits are being adopted, it is crucial for Corporates to understand how to leverage their strengths and how to more importantly leverage collaborations, investments and other forms of partnerships with the more agile and disruptive world of the start-up world.

At the APEC Summit in 2015, during a panel discussion between Jack Ma, founder of Alibaba Group (USD 23+ billion in revenues) and President Barack Obama, Jack Ma said what I believe is to be true: „It is difficult for big companies to keep up with innovation. Innovation always happens outside of the company." Corporates are here to provide financial resources, to provide technical expertise, to provide their unique access to their suppliers and distribution – but they are not here to disrupt. I mean, it would be against your nature to disrupt yourself, no?

Heiko Wöhner: *So, if the corporates cannot be innovative inside, is it natural for the purchasers to look for start-ups?*

Alisée de Tonnac: How natural it is for the Corporate to collaborate outside really depends on where the corporate stands in their strategy towards innovation. You have corporates in the beginning of the reflection; so considering start-ups is not at all natural for every purchaser. When you speak to them you feel like you are speaking a different language. Recently, I was talking to a big corporate and together we were brainstorming the eventuality that maybe their business model would disappear and that, as the middleman, they could be cut out quicker than they thought, and what was interesting was that the biggest psychological barrier was: „No, it has been working for the past decades, so it should work for another decade." Disruption is happening in every single industry and as we are entering the 4th industrial revolution, the speed of current breakthroughs has no historical precedent, as it is not evolving in a linear pace but an exponential one.

This does not only mean the rapidity at which all humans on this planet will be connected but also, what still seems to be the unfathomable impact linked to breakthroughs in fields such as artificial intelligence, robotics, the Internet of Things, 3-D printing, nanotechnology, biotechnology, energy storage, quantum computing and more. And then you have Corporates such as Google or Facebook who are still very young and are still able to keep that agility and define the right partnerships and initiatives to stay close to the talents and entrepreneurs.

Another important element to keep in mind when defining their strategy with regards to innovation is that the implementation of any strategy must be aligned internally from bottom to the top management in the company. Corporates that have been able to leverage the best partnerships with start-ups tend to also have it as a main priority in the Corporate strategy and mission, as a real incentive for not only top but middle management and see it is as a long term play but with a short term emergency to take these small businesses and prototypes seriously.

Heiko Wöhner: *An aspect that is in my mind in that regard is that Corporates feel it is hip or fashionable to have a good contact with start-up companies, to work with them or even to buy them. But I feel that they are not always serious about it. Do you see the two groups of companies: The ones which are seriously looking for business and taking an opportunity and the other ones which just do it because it is hip?*

Alisée de Tonnac: Definitely it is hip. Everyone wants to get a piece of the cake. But not everyone is willing to make/take the necessary measures/risk, and not every company has the right leadership in place to prepare itself for the disruption within their core businesses. Indeed, when it comes to leadership, those that are staying on top are the first to accept and prepare for the disruption of their own industry. Check out Uber that has invested hundreds of millions in driverless cars. These same companies also have a clear and humble understanding that in today's world, in order to keep up they need to work hand in hand with the other innovation spaces such as the start-up world, and don't get me wrong, it does not mean acquiring young trendy start-ups for hundreds of millions that makes the future of the company safe. Indeed assessing the investing in a startup that does not even make revenues.

We see that Corporates, which are clear and realistic positioning, which know their threats and that they can be quickly overturned, have done the best deals. Such companies have understood that it is not about sponsoring start-ups because it is hip but about making serious commercial agreements on every money that enters the start-up ecosystem.

Florian Schupp: *I am also currently thinking and writing about innovation. It is important that innovation has to be used, it has to be marketed because otherwise you will never find out if it can be successful or not. Which organisational changes would you recommend for companies to facilitate this integration of start-ups to enter the ecosystem of start-ups? Is it necessary to have a dedicated department in the company or is it enough if it is done by young and dynamic people in a think tank?*

Alisée de Tonnac: As mentioned, the Corporates' DNA and the start-ups' DNA do not match at all naturally. Therefore, I believe it has to be an external organisation to innovate – but there needs to be a real bridge between the Corporate and the „start-up world". Unfortunately, we have seen examples, where Corporates have launched programmes, but inside their organisations no one was supporting these programs as there were no real incentives and in the end this did not change their bottom line. In those cases it is likely that the start-up dies and only the talents might stay at the Corporate – basically it is a talent acquisition.

This is quite disappointing because we have seen a lot of good products and services, which have simply died because of lack of alignment in the strategic and implementation of such initiatives.

Another important aspect to bridge between the start-up and the Corporate is to incentivise bottom, middle and senior management. Middle management can for example, have a specific bonus system with dedicated time to supporting start-up ventures. The involvement is often very much appreciated in terms of team building and motivation; and by incentivising financially they will also make sure that they commit time and resources.

Florian Schupp: *You answer in two ways: You need an external solution and you need to be able to internalise the ideas with incentives on all levels, with good incentives to activate the people.*

What do you think about our idea from a purchasing perspective: Let's assume you have this organisation that is doing integration and you also incentivise integration. Do you think it is better to fund a start-up, a technological idea or do you think it is also feasible or even better to simply buy from the start up? Because all buyers out there have a luxury which is a burden at the same time: they have to buy. However, this is what we consider people who have created a start-up need the most – and I have also created a start-up during student times, and that is what I felt at least. Is it more important to have funding or to get sales?

Alisée de Tonnac: It always depends on where you stand as an entrepreneur. Sometimes you just do not have a choice. But if you have a choice, I would always go for sales and focus on the monetisation. It's never about just money but about „smart money" where you are able to leverage the network, expertise or business of your investor in order to scale. That is why Corporate investors are sometimes very interesting as they potentially can provide to you customers.

Florian Schupp: *Do you have any examples of companies, which have done very well in connecting with the start-up community?*

Alisée de Tonnac: There are some Corporates, which have developed a good strategy, for example Enel, the utility company in Italy, which is strong in renewable energy. They have launched a business unit for innovation that has for objective to understand the pains in every business unit internally and to find solutions in the outside world – through research with universities but also through different types of interactions with the start-up world. Enel supports these start-ups to grow at a time when they are big enough to have a commercial agreement with the Enel business units

Florian Schupp: *There is the need for right people, the bridge, and the use of good existing infrastructure. You need to be wanting to be in the start-up community.*

Alisée de Tonnac: Exactly. You need to have a clear vision and strategy and a coherent implementation. If you are following the buzzwords and just trending by joining some events it won't get you closer in understanding what the future of your industry is and how your company should position itself. At one point it was about owning its own conference, now it's all about having its own acceleration program or incubation program and who knows what will be trending tomorrow, but if it's not a business decision than it won't go very far.

Heiko Wöhner: *Let's touch the topic of digitalisation itself. Is digitalisation the mean to overcome purchaser's and start-up's cultural and geographical differences? Does it help that people do not need to meet all the time?*

Alisée de Tonnac: I am travelling in 65 markets and 75 cities in our competition because I still believe face-to-face is irreplaceable. What is interesting is that digitalisation is helping the feasibility of many business models, to enhance their success and impact. However, digitalisation does indeed help you to be global from day 1 and it does not come without a doubt that I have team members in all continents. Digitalisation also allows the education of more talents to enter the space and local and international investors to better understand the risks and opportunities.

Florian Schupp: *Let's assume we look into Nigeria for example. Nigeria is one of your target markets. How many of the start-ups in Nigeria are trying to sell a product that is purely driven by data or digitalisation? How many of those are selling a product that is hardware-related?*

Alisée de Tonnac: In my network of the Seedstars competition, where I have around 10,000 applications, it is mostly ICT [information and communication technology] businesses. Why? Our criteria of selection are that a company needs to be less than two years old, have received less than half a million in investment, and have a protypte – a minimum viable product. So usually this takes out high tech businesses, life science or medtech that usually need more time and/or more funding. However, in some regions we see a very strong emphasis on hardware such as in the Easter European region. And hardware has started to become much cheaper now, so we should see the trend continue and grow.

Heiko Wöhner: *Africa is one of the most neglected sourcing markets from a multinational purchasing point of view. How can the enormous potential be opened for* *international supply chains? What should purchasers do to open this resource for their businesses? Purchasers want to find opportunities and to use opportunities.*

Alisée de Tonnac: Like any business looking to expand to a new market, one needs to be very diligent in understanding the „rules of the game" locally. This means understanding the culture (ex: negotiation practices), the history (political risks), the infrastructure (ex: quality of roads or internet infrastructure and penetration), etc. And one needs to accept several sunk costs such as investing in the infrastructure that is lacking in order to reap the benefits of its core business. Take, for example, e-commerces. In Europe, it is much easier to launch one's own e-commerce, whereas in Africa you would need to be willing to put in place your own payment gateway or accept cash on delivery, you would need to integrate rather than outsource departments such as building your own delivery platform. You also need to be willing to invest in educating the local talent. Talent is everywhere, for me this is a fact, but the reality is that they do not have access to the same quality of education, access to the same network or opportunities and this why you need to invest time to train your talent locally so that the business can scale. I don't believe in depending solely on expat. talent. Finally, you need to accept that the one size fits all will probably not work and so you need to prototype a new solution or business model in order to access these markets. That is why so many are talking about micro grid solutions when it comes to producing and distributing electricity for example.

Florian Schupp: *You first have to take the decision to make business in Africa.*

Alisée de Tonnac: Yes, if you are not in it for the long-run, it's not worth investing in Africa.

Florian Schupp: *For purchasers in the Corporates it is often difficult to search in the markets for specific needs. But if they find anything else that might be interesting as well, they do not have the budget and the process to buy it. Like someone going shopping and seeing a very interesting shopping opportunity. Purchasing would benefit from having a credit card for trying out in line with the business model.*

Alisée de Tonnac: I know of many big companies that have given out failure bonuses to employees to really say „It is ok to fail". It is part of the process of innovating. So managers would have a certain amount of budget and could fail up to x. And until this budget they did not need to go through approval. And if you actually failed, you would get a bonus. We also apply a similar principle at Seedstars. Because „the biggest risk in life is not to take risks!"

Heiko Wöhner: *So in summary, you probably say it should be a natural reaction for purchasers to the opportunities in emerging markets to go out and search for the right start-ups in a smart way?*
  Alisée de Tonnac: Absolutely. I wouldn't be doing what I do if I didn't believe in that.

Florian Schupp: *Alisée, thank you very much for the interview.*

Über die Autoren

Dr.-Ing. Florian Schupp ist Einkaufsleiter Automotive und Automotive Aftermarket bei Schaeffler. In Rahmen seiner Dissertation befasste er sich an der Technischen Universität Berlin mit Themen rund um die Strategieentwicklung in Einkauf und Logistik. Dr.-Ing Florian Schupp hat insgesamt 19 Jahre Einkaufserfahrung bei Schaeffler, Continental, Siemens und SONY. Zusammen mit der Universität Lappeenranta, Finnland, und der Universität Catania, Italien, verknüpft er Einkauf und Supply Management mit wissenschaftlicher Forschung. Dr.-Ing. Florian Schupp lehrt an der Technischen Universität Berlin und der Jacobs University Bremen. Seine Forschungsschwerpunkte sind Einkaufsstrategie, Verhaltensaspekte im Einkauf, Integration von Lieferanteninnovation in Parametric-Auctions, Working Capital Management, Buyer-Supplier-Relationship Management und Supply Management. Dr.-Ing Florian Schupp ist Mitglied im Beirat der BLG Logistics, Bremen.

Dr. Heiko Wöhner ist Spezialist Supply Management beim Automobilzulieferer LuK GmbH & Co. KG. Nach seinem Wirtschaftsingenieurstudium in Bremen und Östersund untersuchte er im Rahmen seiner Promotion an der EBS Universität für Wirtschaft und Recht in Wiesbaden, inwiefern Integration mit Kunden und Lieferanten für Unternehmen vorteilhaft ist. Dr. Wöhner sammelte vier Jahre Erfahrungen im Projektmanagement der Bundesvereinigung Logistik (BVL) e. V. und war unter anderem für die inhaltliche Gestaltung des Deutschen Logistik-Kongresses und des Branchenforums Automobil-Logistik mitverantwortlich. Seit 2011 gestaltet Dr. Wöhner die Lieferantenanbindung bei der LuK GmbH & Co. KG und unterstützt Forschung im Bereich Supply Management.

© Daniel Lopez-Paullada

Alisée de Tonnac is CEO & Co-founder of Seedstars World, the global startup competition aiming at impacting people's lives in emerging markets through technology & entrepreneurship.

After traveling for a year around the world to set up the first edition of Seedstars World back in 2013, Alisée is now managing the company and taking it to the next level. The competition is now present in 60+ cities and by next year Seedstars World will be launching 15 strategic hubs (seedspace co-working + seedstars academies) around the world. She now has accumulated deep knowledge of trends in technology, social media & consumer behavior in Emerging Markets.

Born in France, she has spent most of her life abroad, from growing up in Singapore, California, and France to studying and working in Italy and Switzerland and finally accomplishing humanitarian missions in Brazil and Cambodia. As of today, she has traveled in 50+ countries and is currently living in Nigeria.

Graduated from HEC Lausanne and obtained her Master's in International Management at the Bocconi University (Italy) with highest honors. She was also member of the Harvard Model Congress Europe where she won the Award of Excellence. She worked as product manager for luxury brands at L'Oréal Group and was part of the Italian team at Voyage Privé, a leading European startup.

Alisée is a board member of the School of Management of Fribourg. She was nominated Social Entrepreneur Forbes 30 under 30 and Innovation Fellow of Wired UK. She is obsessed with „The Starry Night" by Van Gogh and „Le Violon D'Ingres" by Man Ray.

URL: www.aliseedetonnac.com

Ableitungen zum Einfluss der Digitalisierung auf die Volkswirtschaft

11

Philippe Gillen und Achim Wambach

Zusammenfassung

Die Digitalisierung hat viele Bereiche der Wirtschaft erfasst. Dieses Kapitel beschäftigt sich damit, was das Besondere an der Digitalisierung aus ökonomischer Sicht ist und wie sich Digitalisierung auf die Volkswirtschaft auswirkt. Als erstes wäre die Dynamik zu nennen, durch die sich die Digitalisierung von anderen Entwicklungen wie beispielsweise der Elektrifizierung absetzt. Einhergehende Strukturwandel müssen in kurzer Zeit bewältigt werden, was diese noch einmal verschärft. Die zweite Besonderheit sind die allgegenwärtigen Daten. Immer größere Mengen an Daten werden genutzt, um Prozesse zu optimieren oder gar Individuen zu vermessen. Dies hat Vorteile, wie zum Beispiel effizientere Produktionswege, kann aber auch negative Folgen haben, beispielsweise bei der Versicherung von Individuen. Als dritte Besonderheit ist das verstärkte Aufkommen von Plattformmärkten zu nennen. Ob Plattformen wie eBay oder Airbnb oder Social-Media-Plattformen wie Facebook oder Snapchat – diese gehören zu den größten Gewinnern der digitalen Transformation. Oft werden die Dienste auf einer Seite zu Nullpreisen angeboten, was deren Regulierung für Wettbewerbsbehörden erschwert. Außerdem sind Daten zu einem Zahlungsmittel geworden, als Reaktion hierauf wurde das Gesetz gegen Wettbewerbsbeschränkungen (GWB) novelliert. Ein Puzzle bleibt allerdings: Die zu erwarteten Effizienzgewinne in der Produktion sind in den klassischen Maßen für Produktivität nicht zu sehen.

P. Gillen (✉) · A. Wambach
Mannheim, Deutschland
E-Mail: gillen@zew.de

A. Wambach
E-Mail: wambach@zew.de

© Springer Fachmedien Wiesbaden GmbH 2018
F. Schupp und H. Wöhner (Hrsg.), *Digitalisierung im Einkauf,*
https://doi.org/10.1007/978-3-658-16909-1_11

11.1   Einleitung

Im Jahr 2002 war es das erste Mal möglich, mehr Daten auf digitalen Medien als analog zu speichern (Hilbert und López 2011), dies kann als der Beginn des digitalen Zeitalters gesehen werden. Die ständig steigende Leistung und Miniaturisierung der Mikroprozessoren hat es ermöglicht, diese in fast allen Bereichen einzusetzen. Von der Entwicklung des modernen Internets über den Roboterarm bis hin zum Smartphone haben diese Fortschritte ihren Einzug in das tägliche Leben und die Wirtschaft gefunden, man spricht von der Digitalisierung. Der jederzeit mögliche Zugriff auf alle erdenklichen Informationen oder effizientere Produktionswege sind dabei nur zwei Konsequenzen (BDI und RB 2015). Durch das Internet und die Möglichkeit, über dieses Geld zu verschicken, wurden auch neue Märkte erschlossen, wie zum Beispiel Online Shops oder zahlreiche Plattformmärkte. 2016 haben diese bereits 4 % des deutschen Bruttoinlandproduktes ausgemacht, Tendenz steigend (Statista 2017).

Einzelne digitale Unternehmen sind dabei ungemein erfolgreich – und das in sehr kurzer Zeit. So sind zum Beispiel Start-ups wie Amazon, eBay oder Uber fast über Nacht zu Multi-Milliarden-Dollar-Unternehmen geworden. Dieses Kapitel beschäftigt sich damit, was das Besondere an der Digitalisierung aus ökonomischer Sicht ist und wie sich Digitalisierung auf die Volkswirtschaft auswirkt.

11.2   Die Digitalisierung zeichnet sich aus durch (1.) eine hohe Dynamik, …

Anders als andere technologische Innovationen zeichnen sich die Digitalisierung und die Durchdringung der Wirtschaft mit digitalen Produkten und Anwendungen durch eine hohe Dynamik aus. Die Zeit etwa, die ein Unternehmen heute braucht, um eine Milliarde Dollar Einnahmen zu übersteigen, hat sich immer weiter verkürzt. Während Apple und Dell, gegründet 1976 und 1984, noch acht beziehungsweise neun Jahre gebraucht haben, gelang Amazon dies in fünf Jahren nach ihrer Gründung 1994. Groupon, gegründet 2008, brauchte sogar nur knapp über zwei Jahre. Auch in den sozialen Medien ist die rasante Entwicklung sichtbar. Bei Facebook hat sich im Zeitraum von 2010 bis 2014 die Anzahl der Nutzer in Deutschland verfünffacht. Selbst rein technisch wachsen die Möglichkeiten. Der Intel-Mitbegründer Gordon Moore hatte nach dem Aufkommen der ersten modernen Schaltkreise Anfang der 1960er Jahre erwartet, dass sich die Rechenleistung pro genutztem Raum ca. alle 18 Monate verdoppeln würde (das sog. Moore'sche Gesetz, das nach seiner Formulierung 1965 und einer ersten Korrektur zehn Jahre später auf den heute geläufigen Wert der nämlichen 18 Monate festgelegt wurde). Bisher hat er damit Recht behalten.

Diese Dynamik der digitalen Wirtschaft hat Auswirkungen auf Politik und Gesellschaft, die den damit einhergehenden Strukturwandel gestalten und bewältigen müssen. Bei Entwicklungen wie der Dampfmaschine oder der Elektrifizierung hatte die Industrie Jahrzehnte Zeit, sich auf die neuen Herausforderungen und Gelegenheiten einzustellen. So viel Zeit lässt die Digitalisierung nicht zu.

Durch diese relativ kurzen Zeiträume werden die Probleme, die durch den Strukturwandel entstehen, noch verstärkt. Ein Beispiel für eine Herausforderung, die die Digitalisierung mit sich bringt, ist die Verdrängung von Arbeitsplätzen durch Roboter. Eine Studie von Acemoglu und Restrepo (2017) für das National Bureau of Economic Research zeigt, dass ein zusätzlicher Roboter pro tausend Arbeiter die aggregierte Arbeitslosenquote in den USA um 0,34 Prozentpunkte erhöht und die Löhne um 0,5 Prozentpunkte verringert. In eine Studie des Zentrums für Europäische Wirtschaftsforschung (ZEW) gemeinsam mit der OECD (Arntz et al. 2016) zeichnet sich ab, dass in Deutschland 12 % der Jobs durch die Digitalisierung gefährdet sind. Aber nicht nur Roboter verdrängen Arbeitsplätze, auch sind viele Unternehmen des digitalen Zeitalters weniger angewiesen auf eine große Zahl an Angestellten. Der Kontrast zwischen klassischem und „digitalem" Arbeitgeber wird am Vergleich General Motors und Uber deutlich. Beide Unternehmen haben einen ähnlichen Marktwert von etwa 60 Mrd. USD, allerdings beschäftigt GM weltweit ca. 200.000 Angestellte, während Uber gerade mal 3500 fest angestellte Mitarbeiter hat. Die Fahrer von Uber dagegen arbeiten als Selbstständige. Diese Umstellungen wären sicherlich eine Herausforderung über jeden Zeitraum, die Dynamik der Digitalisierung verschärft diese allerdings noch mal deutlich.

11.3 ..., (2.) allgegenwärtige Daten ...

Daten und die Nutzung von Daten sind charakteristisch für die digitale Ökonomie. Der Zugang zu mehr Daten kann Probleme lösen, die früher nur sehr schwer zu überwinden waren. Ein illustratives Beispiel hierzu liefern Baker und Hubbard (2004). Bevor es möglich war, Lastwagenfahrer über GPS und Telemetrie-Sensoren zu überwachen, waren in der Logistik die Lastwagen häufig im Besitz der Fahrer. Der Hauptgrund hierfür war die Schwierigkeit, in den Verträgen festzuhalten, wie vorsichtig mit dem Lastwagen umgegangen werden sollte und wie er zu pflegen war. Allerdings hatte dies große Nachteile für die Fahrer: Der Fahrer musste die substanziellen Investition für einen Lastwagen tätigen und das damit einhergehende Risiko selbst bewältigen. Mehr Daten haben dieses Dilemma gelöst: Die Firma kann nun fernüberwachen, wie mit dem Wagen umgegangen wird, und kann dies entsprechend vertraglich festhalten. Damit war es dann auch möglich, dass die Spedition den Lastwagen selbst besitzen konnte. Dieses Beispiel zeigt, wie der Zugang zu mehr und besseren Daten durch die Digitalisierung Informationsasymmetrien reduzieren kann.

11.3.1 Big data als neues Werkzeug der Digitalisierung

Neben den Datenmengen haben auch neue Analyseverfahren Einzug in die Unternehmen genommen, wie zum Beispiel big data. Big data beschreibt die Auswertung großer Datenmengen aus verschiedenen Quellen in hoher Geschwindigkeit. So fließen im Unternehmen der Industrie 4.0 neben üblichen Kundendaten zum Beispiel Sensordaten von Maschinen oder Geräten, Lagerdaten oder auch Daten wie Wetterprognosen in die Analysen ein.

Eine Studie des ZEW (Astor et al. 2016) hat ergeben, dass Big-data-Analysen mit 44 % vornehmlich in den Finanz- und Versicherungsdienstleistungen genutzt werden, weitere Branchen mit einer relativ hohen Anwendungsquote sind mit 25 % der Fahrzeugbau sowie Chemie- und Pharmaunternehmen mit 24 %. Die Unternehmen verwenden die verfügbare Datenflut, um ihre Produkte zu verbessern und um die Interessen ihrer Kunden besser zu verstehen. Im Bereich der Verbrauchsgüter spielt big data mit elf Prozent eine deutlich geringere Rolle.

Big-data-Analysen können den Unternehmen Entscheidungsgrundlagen liefern, zum Beispiel im Risikomanagement der Versicherungen oder auch im Produktions- und Supply Chain Management. Durch diese zusätzlichen Informationen können Unternehmen die Produktivität erhöhen und neue Geschäftsmodelle entwickeln. So entstehen derzeit viele Start-ups die auf Kombination von big data mit anderen Technologien wie Cloud Computing setzen.

11.3.2 Vermessung des Individuums

Insbesondere die nutzerbezogenen Daten sind für die Marktprozesse relevant. Sie ermöglichen eine „Vermessung" des Individuums. So werden beim Surfen im Internet von diversen Unternehmen Daten über das Verhalten und die Interessen im Netz gesammelt. Diese können dann mit Datenbanken mit personenbezogenen Daten wie zum Beispiel Alter, Geschlecht oder Wohnort abgeglichen werden. Nach und nach wird damit ein Profil des Nutzers erstellt, welches für individuelle Angebote und gezielte Werbung genutzt wird. Diese Form der Diskriminierung kann je nach Sektor positive oder negative Auswirkungen haben. Durch die Nutzung der Daten können gezielt Produkte beworben werden, die den Nutzer potenziell interessieren. Auf der anderen Seite bereitet die Klassifizierung der Individuen etwa im Versicherungsbereich Sorge, wenn sich beispielsweise Personen wegen ihrer Datenhistorie nicht mehr versichern können.

Die Relevanz nutzerbezogener Daten für Unternehmen zeigt sich exemplarisch an einer ZEW-Studie (Kummer und Schulte 2016) zum Thema Apps für Smartphones, die eine Austauschbeziehung zwischen Preisen und Privatsphäre feststellt. Kostenpflichtige Apps haben laut der Untersuchung bessere Datenschutzbestimmungen als kostenlose Apps. Daten werden dabei zu Zahlungsmitteln und ersetzen dadurch an manchen Stellen Geld als zentrales Allokationsinstrument.

11.4   ... und (3.) Plattformmärkte

Die dritte ökonomische Besonderheit sind Plattformmärkte, die bezeichnend für die digitale Wirtschaft sind. Plattformen wie eBay oder Airbnb, die nicht direkt ein Produkt oder eine Dienstleistung verkaufen, sondern deren Geschäftsmodell darin besteht, Angebot und Nachfrage zusammenzubringen, gehören zu den größten Gewinnern der digitalen Transformation. Aber auch Plattformen wie Facebook, Snapchat oder Twitter sind sehr erfolgreich: Der Wert des Social-Media-Giganten Facebook wird auf ca. 400 Mrd. USD geschätzt. Diese Entwicklung ist nicht unproblematisch. Einerseits haben besonders soziale Plattformen durch direkte und indirekte Netzwerkeffektive eine natürliche Tendenz zu Monopolen, ein „digitaler Mittelstand" existiert kaum. Hierfür sorgen zusätzlich Pioniermarkteffekte und die geringen Kosten für Kommunikation. WhatsApp und Facebook Messenger zum Beispiel, die beide zu Facebook gehören, haben den Trend zu Online Messaging früh erkannt und teilen sich heute fast 80 % des Marktes in Deutschland (Kroker 2016). Aber auch die zuvor genannten Netzwerkeffekte machen Wettbewerb schwierig. Schafft es eine App oder ein Dienst, eine kritische Masse zu erreichen, ist die Eintrittsbarriere für neue Wettbewerber hoch. Multihoming, die Möglichkeit der gleichzeitigen Nutzung mehrerer Plattformen, wirkt dem entgegen.

Bezeichnend für die Plattformökonomie ist auch, dass es viele Dienste zu Nullpreisen gibt. Dies ist eine häufig zu beobachtende Preissetzung bei mehrseitigen Plattformen, bei denen eine Plattformseite (etwa der Endkonsument) subventioniert wird und die Einnahmen auf der anderen Plattformseite (etwa bei den Werbetreibenden) generiert werden.

Die Wettbewerbsbehörden haben sich darauf eingestellt, dass nun Güter ver-
schenkt oder gegen die Nutzung von Daten getauscht werden. Ein wichtiger Schritt
dahin ist die Anpassung des Wettbewerbsrechts in Form der 9. Novelle des Geset-
zes gegen Wettbewerbsbeschränkungen (GWB). Hier wurde konkret der Frage
nachgegangen, ob ein Markt vorliegt, wenn eine Leistung unentgeltlich, also ohne
monetäre Transaktion erbracht wird. In der vom Bundestag im März 2017 beschlos-
senen Novelle wurde festgehalten: „Der Annahme eines Marktes steht nicht entgegen, dass eine Leistung unentgeltlich erbracht wird." Und weiter: „Insbesondere bei
mehrseitigen Märkten und Netzwerken sind bei der Bewertung der Marktstellung
eines Unternehmens auch zu berücksichtigen: […] direkte und indirekte Netzwerkeffekte, […] sein Zugang zu wettbewerbsrelevanten Daten […]."

11.5 Produktivitätsschub bleibt aus

Die Digitalisierung hat viele Bereiche der Wirtschaft erfasst. Bei allen genannten Vorteilen und trotz der angedeuteten Schwierigkeiten würde man erwarten,
dass sich dies auch in den volkswirtschaftlichen Größen widerspiegelt. Allerdings
sieht die Datenlage anders aus: Seit einigen Jahren sinkt das Wachstum der Produktivität in Europa und auch im Rest der Welt. Man spricht in diesem Zusammenhang vom „Productivity Puzzle".

„Produktivität ist nicht alles, aber auf lange Sicht fast alles", so der US-ame-
rikanische Ökonom Paul Krugman. Die Zu- oder Abnahme des Wohlstandes in
einem Land hängt direkt von der Produktivität ab. Das Wachstum der Produktivität ist maßgeblich dafür, ob Reallohnerhöhungen möglich sind oder nicht. Um
noch einmal Paul Krugman zu zitieren „Die Fähigkeit eines Landes, den Lebens-
standard zu erhöhen, hängt fast ausschließlich davon ab, um wie viel es die Produktion per Arbeiter steigern kann."

Von der digitalen Revolution sollte erwartet werden, dass sie die durchschnitt-
liche Arbeitsproduktivität in den Industrieländern massiv erhöht – so wie es vor
200 Jahren die Dampfmaschine tat und vor 100 Jahren die Elektrifizierung. Doch
ganz im Gegenteil, das Wachstum der Produktivität ist in den letzten 30 Jahre
stetig gesunken (siehe Abb. 11.1). Während in den 1970er Jahren das Wachstum
der Produktivität in Deutschland im Jahresdurchschnitt noch bei vier Prozent lag,
hatte es sich in den 1980er und 1990er Jahren bereits halbiert – auf zwei Prozent.
Danach sank es weiter auf 1,1 % pro Jahr in den 2000er Jahren, um dann den ver-
gangenen vier Jahren sogar nur noch 0,5 % zu erreichen.

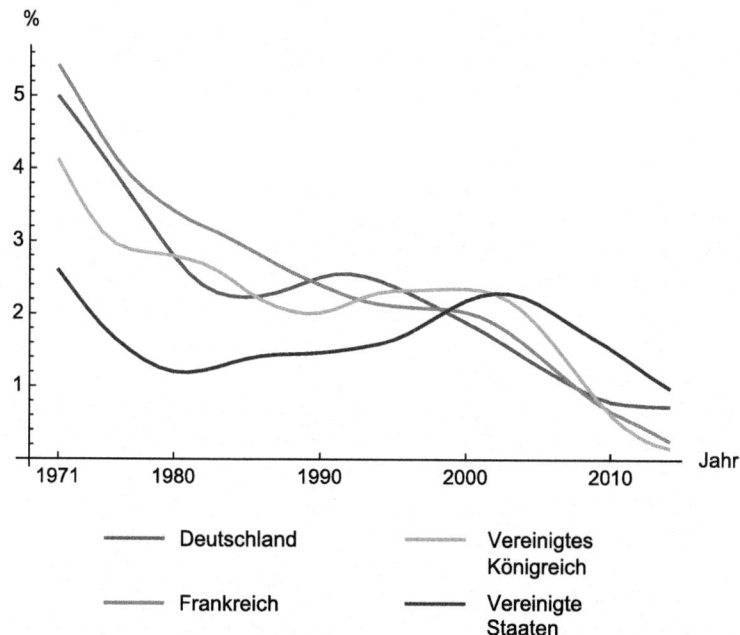

Abb. 11.1 Arbeitsproduktivität je Erwerbstätigenstunde in ausgewählten Ländern: Trendwachstum. (Quelle: Eigene Darstellung; Daten: SVR Jahresgutachten 15/16) (Online farbig)

Ist die Digitalisierung also doch nicht der nächste große Schritt in der industriellen Entwicklung? Oder gibt es andere Gründe für die fehlenden Steigerungen der Produktivität? Mehrere Erklärungen werden vorgebracht.

Ein erster Grund könnten Nachwirkungen von Finanz- und Euro-Krise sein. Unternehmen, die sich von dieser Krise erholt haben, scheuen laut den Verfechtern dieser Hypothese davor zurück, größere Investitionen im Digitalbereich wie Server oder Robotik zu tätigen. Nach der Krise sind viele Unternehmen dazu übergegangen, lieber Geldreserven anzulegen oder Aktien zurückzukaufen. Außerdem beklagen viele Unternehmen, dass es seit der Krise nicht mehr genug Nachfrage nach ihren Produkten gibt, um große Investitionen zu rechtfertigen (Baumann und Vasardani 2016; SVR 2015, S. 303 ff.).

Eine zweite mögliche Erklärung ist, dass die aktuell verwendeten Maße für Produktivität veraltet sind und nicht mehr Schritt halten können mit den vielen neuen Produkten und Dienstleistungen, die die Digitalisierung mit sich bringt. So schreibt Martin Feldstein, Professor für Ökonomie in Harvard: „Diejenigen, die pessimistisch sind was den Wirtschaftswachstum angeht, liegen falsch. Diese Statistiken unterschätzen das tastsächliche Wachstum des BIP, der Produktivität und der Haushaltseinkommen" (Feldstein 2015). Der Sachverständigenrat schreibt hierzu, dass die Probleme bei der Datenerfassung besonders im Dienstleistungsbereich bestehen (Sachverständigenrat zur Begutachtung der Gesamtwirtschaftlichen Entwicklung [SVR], Zukunftsfähigkeit in den Mittelpunkt. Jahresgutachten 2015, Textziffer 618). Oft wird die Wertschöpfung hier über die Kosten- und Lohnseite bestimmt, und Zu- oder Abnahmen in der Qualität sind damit nur schwer zu messen. Eine hedonische Preisbereinigung, also unter Berücksichtigung von Qualität, sei kaum möglich, und es ist daher denkbar, dass die Wertschöpfung bei den Dienstleistungen unterzeichnet wird.

Kritiker dieser These liefern allerdings zwei Gegenargumente, warum diese Erklärung die sinkende Produktivität nicht gänzlich beschreiben kann (Byrne et al. 2016). Einerseits hat es diese Fehler bei der Vermessung in der Informations- und Kommunikationstechnik schon immer gegeben, so auch zum Beispiel in den USA von 1995 bis 2000, als die Produktivität dort allerdings gestiegen ist. Andererseits hat sich der Rückgang nicht auf die Dienstleistungen beschränkt, sondern ist auch in den einfacher messbaren Sektoren sichtbar.

Die dritte Möglichkeit besteht, wie oben angedeutet, darin, dass technischer Fortschritt heute keinen so großen Einfluss mehr auf die Ökonomie hat, wie es früher der Fall war. Robert Gordon, Professor für Ökonomie an der Northwestern University, schreibt, dass während Smartphone und das Internet zwar ohne Zweifel große Errungenschaften menschlichen Einfallsreichtums seien, diese jedoch nicht vergleichbar wären mit dem Einfluss des Verbrennungsmotors, Stroms oder sogar Flugzeugen (Gordon 2012, S. 14). Gordons Annahme kulminiert in dem Gedankenexperiment: „Auf was würden Sie lieber verzichten, ihr Badezimmer mit Toilette in der Wohnung oder auf ihr iPhone?" (ebd.) Sollte Gordon Recht haben, ist das langsame Wachstum der Produktivität kein temporärer Trend, sondern ein permanenter Zustand, auf den sich eingestellt werden muss.

11.6   Fazit

Die Digitalisierung stellt die Wirtschaft vor große Herausforderungen. Die Behörden haben darauf reagiert. So wurde in der Wettbewerbspolitik das GWB novelliert, um besser auf die Besonderheiten der digitalen Wirtschaft eingehen zu können. Mit der EU-Datenschutzgrundverordnung, die 2018 in Kraft treten soll, wurde ein wichtiger Schritt im Verbraucherschutz getätigt. Der richtige Umgang mit der Sharing Economy (etwa Uber und Airbnb), den „Over-the-top-playern" im Telekommunikationssektor (etwa Skype) oder den neuen Fintech-Unternehmen im Finanzsektor wird die Behörden auch in den nächsten Jahren intensiv beschäftigen.

Innovationen der Digitalisierung wie das Internet oder Smartphones haben die subjektive Lebensqualität von vielen verbessert. Trotzdem sind diese Veränderungen noch nicht in den Daten zur Produktivität messbar. Die Ursachenforschung für dieses Puzzle bleibt eine spannende Frage.

Literatur

Acemoglu, D., Restrepo, P.: Robots and Jobs: Evidence from US Labor Markets (2017)

Arntz, M., Gregory, T., Zierahn, U.: The risk of automation for jobs in OECD countries: a comparative analysis. OECD Social, Employment, and Migration Working Papers, **189** (2016)

Astor, M., et al.: Endbericht: Innovativer Mittelstand 2015 – Herausforderungen. Trends und Handlungsempfehlungen für Wirtschaft und Politik, ZEW-Gutachten und Forschungsberichte (2016)

Baker, G.P., N. Hubbard, T.: Contractibility and asset ownership: on-board computers and governance in US trucking. Q. J. Econ. **119**(4), 1443–1479 (2004)

Baumann, U., Vasardani, M.: The slowdown in US productivity growth – what explains it and will it persist? Bank of Greece Working paper No. 215 (2016)

BDI, Berger, R.: Die Digitale Transformation der Industrie – Eine europäische Studie von Roland Berger Strategy Consultants im Auftrag des BDI. München, Berlin (2015)

Byrne, D.M., Fernald, J.G., Reinsdorf, M.B.: Does the United States have a productivity slowdown or a measurement problem? Brookings Pap. Econ. Act. **2016**(1), 109–182 (2016)

Feldstein, M.: The US Underestimates Growth. In: Wall Street Journal, 18.05.2015 (2015)

Gordon, R.J.: Is US economic growth over? Faltering innovation confronts the six headwinds. National Bureau of Economic Research, Working Paper, No. 18315. http://www.nber.org/papers/w18315.pdf. Zugegriffen 30 Juni 2017 (2012)

Hilbert, M., López, P.: The world's technological capacity to store, communicate, and compute information. Science **332**(6025), 60–65 (2011)

Kroker, M.: Facebook Messenger mit 800 Millionen Nutzern; 23 Prozent Marktanteil in Deutschland. http://blog.wiwo.de/look-at-it/2016/01/13/facebook-messenger-mit-800-millionen-nutzern-23-prozent-marktanteil-in-deutschland/. Zugegriffen 26 Juni 2017 (2016)

Kummer, M.E., Schulte, P.: When private information settles the bill: money and privacy in Google's market for smartphone applications, S. 16–031. Zentrum für Europäische Wirtschaftsforschung, ZEW Discussion Paper No (2016)

Sachverständigenrat zur Begutachtung der Gesamtwirtschaftlichen Entwicklung (SVR), Zukunftsfähigkeit in den Mittelpunkt. Jahresgutachten 2015/16 (2015)

Statista.   https://www.statista.com/statistics/250703/forecast-of-internet-economy-as-percentage-of-gdp-in-g-20-countries/. Zugegriffen 26 Juni 2017 (2017)

Über die Autoren

Philippe Gillen ist Mitarbeiter am Zentrum für Europäische Wirtschaftsforschung in der Forschungsgruppe „Marktdesign". Seine Forschungsinteressen umfassen alle Gebiete des Marktdesigns mit Schwerpunkten in der angewandten Auktionstheorie und im Einkaufsdesign.

Prof. Achim Wambach, PhD, ist seit April 2016 Präsident des Zentrums für Europäische Wirtschaftsforschung (ZEW), Mannheim. Vor seinem Wechsel an das ZEW war er Direktor des Instituts für Wirtschaftspolitik (iwp) an der Universität zu Köln und Co-Direktor des Europäischen Zentrums für freie Berufe der Universität zu Köln. 2014 wurde Achim Wambach zum Mitglied der Monopolkommission ernannt und im März 2016 zum Vorsitzenden gewählt. Für die Zeit von Februar bis September 2017 wurde er in den Wissenschaftlichen Beirat zur Weiterentwicklung des Risikostrukturausgleichs beim Bundesversicherungsamt berufen. Er ist Mitglied des Wissenschaftlichen Beirats beim Bundesministerium für Wirtschaft und Energie. In seiner Forschung beschäftigt sich Achim Wambach unter anderem mit Fragen der Wettbewerbspolitik sowie Informationsproblemen auf Märkten. Dabei arbeitet er mit Methoden des Marktdesigns und der Industrieökonomik.

Stichwortverzeichnis

© Springer Fachmedien Wiesbaden GmbH 2018
F. Schupp und H. Wöhner (Hrsg.), *Digitalisierung im Einkauf,*
https://doi.org/10.1007/978-3-658-16909-1

Printed by Printforce, the Netherlands